Ultrasound in Chemistry

Edited by
José-Luis Capelo-Martínez

Further Reading

Loupy, A.

Microwaves in Organic Synthesis

Second, Completely Revised and Enlarged Edition

2006
Hardcover
ISBN: 978-3-527-31452-2

Kappe, C. O., Dallinger, D., Murphree, S. S.

Practical Microwave Synthesis for Organic Chemists

Strategies, Instruments, and Protocols

2009
Hardcover
ISBN: 978-3-527-32097-4

Koch, M. V., VandenBusche, K. M., Chrisman, R. M.

Microinstrumentation

for High Throughput Experimentation and Process Intensification

2007
Hardcover
ISBN: 978-3-527-31425-6

Kromidas, S., Kuss, H.-J.

Quantification in LC and GC

A Practical Guide to Good Chromatographic Data

2009
Hardcover
ISBN: 978-3-527-32301-2

Ultrasound in Chemistry

Analytical Applications

Edited by
José-Luis Capelo-Martínez

WILEY-VCH Verlag GmbH & Co. KGaA

The Editors

Prof. J. L. Capelo-Martínez
Dept. de Química-Requimte
Quinta da Torre
2829-516 Caparica
Portugal

All books published by Wiley-VCH are carefully produced. Nevertheless, authors, editors, and publisher do not warrant the information contained in these books, including this book, to be free of errors. Readers are advised to keep in mind that statements, data, illustrations, procedural details or other items may inadvertently be inaccurate.

Library of Congress Card No.: applied for

British Library Cataloguing-in-Publication Data
A catalogue record for this book is available from the British Library.

Bibliographic information published by the Deutsche Nationalbibliothek
The Deutsche Nationalbibliothek lists this publication in the Deutsche Nationalbibliografie; detailed bibliographic data are available on the Internet at http://dnb.d-nb.de.

© 2009 WILEY-VCH Verlag GmbH & Co. KGaA, Weinheim

All rights reserved (including those of translation into other languages). No part of this book may be reproduced in any form – by photoprinting, microfilm, or any other means – nor transmitted or translated into a machine language without written permission from the publishers. Registered names, trademarks, etc. used in this book, even when not specifically marked as such, are not to be considered unprotected by law.

Typesetting   Thomson Digital, Noida, India
Printing   betz-druck GmbH, Darmstadt
Binding   Litges & Dopf Buchbinderei GmbH, Heppenheim
Cover Design   Adam-Design, Bernd Adam, Weinheim

Printed in the Federal Republic of Germany
Printed on acid-free paper

ISBN: 978-3-527-31934-3

Contents

Preface *XI*
List of Contributors *XIII*

1   **The Power of Ultrasound** *1*
Hugo Miguel Santos, Carlos Lodeiro, and José-Luis Capelo-Martínez
1.1   Introduction *1*
1.2   Cavitation *1*
1.2.1   Parameters Affecting Ultrasonic Cavitation *3*
1.2.1.1   Frequency *3*
1.2.1.2   Intensity *4*
1.2.1.3   Solvent *5*
1.2.1.4   Temperature *5*
1.2.1.5   External Pressure and Bubbled Gas *5*
1.2.1.6   Direct and Indirect Ultrasonic Application *6*
1.3   Common Ultrasonic Devices Used in Analytical Chemistry *6*
1.3.1   Ultrasonic Bath *6*
1.3.1.1   Types of Ultrasonic Baths *7*
1.3.1.2   How to Locate the Most Efficient Place Inside an Ultrasonic Bath *8*
1.3.1.3   Temperature Control *8*
1.3.1.4   Shape and Material of Reaction Container *9*
1.3.2   The Ultrasonic Probe *9*
1.3.2.1   Parts of an Ultrasonic Probe *10*
1.3.2.2   Types of Ultrasonic Probes *10*
1.3.2.3   Dead Zones *11*
1.3.2.4   How to Choose the Correct Ultrasonic Probe *12*
1.3.2.5   Temperature Control *13*
1.3.2.6   Shape and Material of Reaction Container *14*
1.4   Current Ultrasonic Devices for New Analytical Applications *14*
     References *15*

Ultrasound in Chemistry: Analytical Applications. Edited by José-Luis Capelo-Martínez
Copyright © 2009 WILEY-VCH Verlag GmbH & Co. KGaA, Weinheim
ISBN: 978-3-527-31934-3

2	**Ultrasonic Energy as a Tool for Sample Treatment for the Analysis of Elements and Elemental Speciation** *17*
	Hugo Miguel Santos, Carlos Lodeiro, and José-Luis Capelo-Martínez
2.1	Introduction *17*
2.2	Parameters Influencing Element Ultrasonic Solid–Liquid Extraction *17*
2.2.1	Extracting Reagent *17*
2.2.1.1	Extracting Reagents for Total Element Extraction *18*
2.2.1.2	Extracting Reagents for Elemental Speciation *19*
2.2.1.3	Extracting Reagents for Sequential Extraction Schemes *19*
2.2.2	Matrix Properties *20*
2.2.2.1	Type of Matrix *20*
2.2.2.2	Mass of Matrix *20*
2.2.2.3	Sample Size *22*
2.2.3	Ultrasonic Device *22*
2.2.3.1	Type of Ultrasonic Device *22*
2.2.3.2	Time of Ultrasonication *23*
2.2.3.3	Ultrasonic Amplitude *23*
2.2.3.4	Ultrasonic Frequency *23*
2.2.3.5	Temperature of Sonication *23*
2.3	US-SLE from Soils and Sediments *24*
2.4	US-SLE from Sewage Sludge *24*
2.5	US-SLE Extraction from Plants *24*
2.6	Extraction from Soft Tissues *27*
2.7	Total Element Determination *27*
2.7.1	US-SLE and US-SS for F-AAS *27*
2.7.2	US-SLE and US-SS for ET-AAS *28*
2.7.3	US-SLE and US-SS for CV and HG Employed with AAS or AFS *28*
2.8	Elemental Fractionation and Elemental Speciation *30*
2.8.1	What is Speciation? *30*
2.8.2	Shortening Sequential Fractionation Schemes *31*
2.8.3	Speciation for Soils and Sediments *34*
2.8.4	Speciation from Plants *34*
2.8.5	Speciation from Soft Tissues *36*
2.8.6	Speciation from Other Types of Samples *43*
2.9	On-Line Applications *45*
2.9.1	Open and Closed Systems *46*
2.9.2	UB *47*
2.9.3	UP *47*
2.10	Current Trends *48*
2.10.1	Accelerating Liquid–Liquid Extractions *48*
2.10.2	Chemical Vapor Formation *49*
2.11	Conclusion *49*
	References *50*

3	**Ultrasonic Assisted Extraction for the Analysis of Organic Compounds by Chromatographic Techniques** 55
	Raquel Rial-Otero
3.1	Introduction 55
3.2	Overview of Classic and Modern Extraction Procedures for Organics 56
3.3	Ultrasonic Assisted Extraction (UAE) 60
3.3.1	Basic Principles 60
3.3.2	Parameters Influencing Ultrasonic Assisted Extraction 61
3.3.2.1	Amount of Sample 61
3.3.2.2	Sample Particle Size 61
3.3.2.3	Extraction Solvent 61
3.3.2.4	pH of Extracting Solution 62
3.3.2.5	Solvent Volume 62
3.3.2.6	Sonic Power 62
3.3.2.7	Frequency 63
3.3.2.8	Extraction Time 63
3.3.2.9	Extraction Temperature 63
3.3.3	Applications 63
3.3.3.1	Liquid Samples 64
3.3.3.2	Solid Samples 64
3.3.3.3	Clean-Up 70
3.4	Coupling Ultrasound with Other Extraction Techniques 71
3.4.1	Coupling Solid Phase Microextraction (SPME) and Ultrasound 71
3.4.1.1	Improving the Extraction Procedure in Direct-SPME 71
3.4.1.2	Improving the Extraction Procedure in HS-SPME 73
3.4.1.3	Facilitating the Desorption Process 73
3.4.2	Coupling Stir Bar Sorptive Extraction (SBSE) and Ultrasound 74
3.5	Comparison between UAE and Other Extraction Techniques 75
3.6	Conclusion 76
	References 77
4	**Electrochemical Applications of Power Ultrasound** 81
	Neil Vaughan Rees and Richard Guy Compton
4.1	Introduction 81
4.2	Electrochemical Cell and Experimental Setup 87
4.3	Voltammetry Under Insonation 87
4.4	Trace Detection by Stripping Voltammetry 88
4.4.1	Classical Electroanalysis 89
4.4.2	Electroanalysis Facilitated by Ultrasound 90
4.4.3	Applications of Sono-Anodic Stripping Voltammetry (Sono-ASV) 90
4.5	Biphasic Sonoelectroanalysis 90
4.5.1	Determination of Lead in Petrol 90
4.5.2	Extraction and Determination of Vanillin 92
4.5.3	Detection of Copper in Blood 92
4.6	Microelectrodes and Ultrasound 93

4.6.1	Insights into Bubble Dynamics	*93*
4.6.2	Measurement of Potentials of Zero Charge (PZC)	*95*
4.6.3	Particle Impact Experiments	*96*
4.7	Conclusion	*102*
	References	*103*

5 Power Ultrasound Meets Protemics *107*
Hugo Miguel Santos, Carlos Lodeiro, and José-Luis Capelo-Martínez

5.1	Introduction	*107*
5.2	Protein Identification through Mass-Based Spectrometry Techniques and Peptide Mass Fingerprint	*108*
5.3	Classic In-Gel Protein Sample Treatment for Protein Identification through Peptide Mass Fingerprint	*108*
5.4	Ultrasonic Energy for the Acceleration of In-Gel Protein Sample Treatment for Protein Identification through Peptide Mass Fingerprint	*111*
5.4.1	Washing, Reduction and Alkylation Steps	*111*
5.4.2	In-Gel Protein Digestion Process	*113*
5.4.2.1	Sample Handling	*113*
5.4.2.2	Sonication Volume	*115*
5.4.2.3	Sonication Time	*115*
5.4.2.4	Sonication Amplitude	*115*
5.4.2.5	Protein to Trypsin Ratio	*115*
5.4.2.6	Temperature	*116*
5.4.2.7	Solvent	*116*
5.4.2.8	Minimum Amount of Protein Identified	*116*
5.4.2.9	Reduction and Alkylation Steps	*117*
5.4.2.10	Comparison with Other Types of Rapid Sample Treatments	*117*
5.4.2.11	Influence of Protein Staining	*117*
5.5	Classic In-Solution Protein Sample Treatment for Protein Identification through Peptide Mass Fingerprint	*118*
5.6	Ultrasonic Energy for the Acceleration of the In-Solution Protein Sample Treatment for Protein Identification through Peptide Mass Fingerprint	*121*
5.6.1	In-Solution Protein Denaturation	*121*
5.6.2	In-Solution Protein Reduction and Alkylation	*122*
5.6.3	In-Solution Protein Digestion	*124*
5.6.4	Clean In-Solution Protein Digestion	*124*
5.7	Conclusion	*125*
	References	*126*

6 Beyond Analytical Chemistry *129*
Carlos Lodeiro and José-Luis Capelo-Martínez

6.1	Introduction	*129*
6.2	Sonochemistry for Organic Synthesis	*129*

6.3	Ultrasonic Enhanced Synthesis of Inorganic Nanomaterials	*137*
6.4	Sonochemistry Applied to Polymer Science	*139*
6.4.1	Introduction to Polymers *139*	
6.4.2	Ultrasonication in Sample Treatment for Polymer Characterization *140*	
6.4.2.1	Introduction to Polymer Characterization	*140*
6.4.2.2	Overview of Sample Preparation for MALDI Analysis of Polymers	*141*
6.4.2.3	Ultrasonic Energy as a Tool for Fast Sample Treatment for the Characterization of Polymers by MALDI	*142*
6.4.3	Ultrasonic-Induced Polymer Degradation for Polymer Characterization	*145*
6.4.4	Ultrasonication for the Preparation of Imprinted Polymers	*145*
6.5	Conclusion *148*	
	References *148*	

Index *151*

Preface

The main reason for writing this book was the desire to promote the use of ultrasound as a tool for sample treatment more widely among the Analytical Chemistry community. This text is intended to serve as a laboratory guide in that it addresses the more practical aspects of the subject, reflecting the most important applications of ultrasound for sample treatment reported in the literature to date. This is definitely not a theoretical text.

Chapter 1 reviews the ultrasonic devices available nowadays to perform sample treatment for chemical analysis. Chapter 2 is devoted to applications of ultrasound for the extraction and determination of elements, including element speciation, in a wide variety of matrixes, whilst Chapter 3 is dedicated to the extraction and analysis of organic compounds. Chapter 4 covers the different uses of ultrasound in analytical electrochemistry. Chapter 5 is dedicated to the acceleration of sample treatments of one growing area in analytical chemistry: proteomics – this is one of the most recent applications of ultrasonic energy. Finally, Chapter 6 deals with further applications of ultrasonic energy, not only in analytical chemistry but in the general chemistry arena.

I thank Dr Heike Nöthe and Dr Manfred Köhl from the publishers, Wiley, who kindly proposed me the edition and writing of this book. I am also grateful to all the co-authors, Professor R.G. Compton, Dr N.V. Rees, Dr C. Lodeiro, Dr R. Rial-Otero and H.M. Santos, M.Sc.

Finally, we hope that you will find this text useful and that ultrasonic energy will become not only a routine laboratory procedure but also an exciting research field for analytical chemists.

Caparica, Portugal
October 2008

José-Luis Capelo-Martínez

List of Contributors

José-Luis Capelo-Martínez
Universidade Nova de Lisboa
REQUIMTE
Departamento de Química
Faculdade de Ciências e Tecnología
2829-516 Monte de Caparica
Portugal

Richard Guy Compton
Oxford University
Physical & Theoretical Chemistry
Laboratory
South Parks Road
Oxford OX1 3QZ
United Kingdom

Carlos Lodeiro
Universidade Nova de Lisboa
REQUIMTE
Departamento de Química
Faculdade de Ciências e Tecnología
2829-516 Monte de Caparica
Portugal

Neil Vaughan Rees
Oxford University
Physical & Theoretical Chemistry
Laboratory
South Parks Road
Oxford OX1 3QZ
United Kingdom

Raquel Rial-Otero
Universidade Nova de Lisboa
REQUIMTE
Departamento de Química
Faculdade de Ciências e Tecnología
2829-516 Monte de Caparica
Portugal

Hugo Miguel Santos
Universidade Nova de Lisboa
REQUIMTE
Departamento de Química
Faculdade de Ciências e Tecnología
2829-516 Monte de Caparica
Portugal

1
The Power of Ultrasound

Hugo Miguel Santos, Carlos Lodeiro, and José-Luis Capelo-Martínez

1.1
Introduction

During the last 20 years we have witnessed an amazing increase in the application of ultrasonic energy in different fields of science. This is especially true for analytical chemistry. The number of manuscripts devoted to almost all kinds of analysis dealing with the uses of ultrasonic energy continues to grow year by year. As the uses of ultrasonication have become increasingly important in analytical chemistry so to has the importance of the type of ultrasonic device chosen to work with. Figure 1.1 shows the most common ultrasonic devices used nowadays in analytical applications. Not all devices perform equally and neither are all intended for the same applications. Therefore, the first thing to acquire when developing analytical chemistry with the aid of ultrasonication is a knowledge of the differences among the ultrasonic apparatus available, especially of the advantages and disadvantages expected for each one. Therefore, this chapter explains the state-of-the-art of ultrasonic technology as applied to analytical chemistry.

1.2
Cavitation

Sound, including ultrasound, is transmitted through any physical medium by waves that compress and stretch the molecular spacing of the medium through which it passes. As the ultrasound cross the medium (Figure 1.2) the average distance between the molecules will vary as they oscillate about their mean position. When the negative pressure caused for an ultrasonic wave crossing a liquid is large enough, the distance between the molecules of the liquid exceeds the minimum molecular distance required to hold the liquid intact, and then the liquid breaks down and voids are created. Those voids are the so-called cavitation bubbles [1–3].

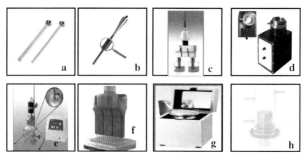

Figure 1.1 Advances in ultrasonic technology: (a) silica glass probe; (b) spiral probe; (c) dual probe; (d) sonoreactor; (e) and (f) multi probe; (g) microplate horns; (h) cup horns. Parts (a,b,e,f and h) are reproduced with permission of the Bandelin company; part (d) is reproduced with permission of the Dr Hielscher company; (c) and (g) are reproduced with permission of Misonix company. Adapted from Ref. [13].

As the liquid compresses and stretches, the cavitation bubbles can behave in two ways [1]. In the first, called stable cavitation, bubbles formed at fairly low ultrasonic intensities (1–3 W cm^{-2}) oscillate about some equilibrium size for many acoustic cycles. In the second, called transient cavitation, bubbles are formed using sound intensities in excess of 10 W cm^{-2}. Transient bubbles expand through a few acoustic cycles to a radius of at least twice their initial size, before collapsing violently on compression (Figure 1.2).

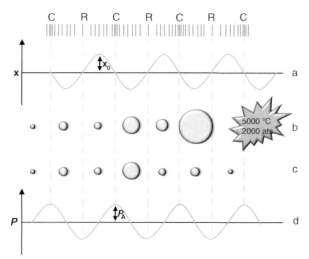

Figure 1.2 Creation of stable cavitation bubbles and creation and collapse of transient and stable cavitation bubbles.
(a) Displacement (x) graph; (b) transient cavitation; (c) stable cavitation; (d) pressure (P) graph.

Transient bubble collapsing is considered to be the main source of the chemical and mechanical effects of ultrasonic energy. Each collapsing bubble can be considered as a microreactor in which temperatures of several thousands degrees and pressures higher than one thousand atmospheres are created instantaneously [4].

From an analytical chemistry point of view, most of the effects of interest regarding ultrasonication are related to cavitation. Cavitation causes solute thermolysis along with the formation of highly reactive radicals and reagents, such as hydroxyl radicals and hydrogen peroxide, which induce drastic reactive conditions in the liquid media [3]. In addition, if a solid is present in solution, the sample size of the particles is diminished by solid disruption, thereby increasing the total solid surface in contact with the solvent. In this way, ultrasonication remains unique, since no other method of sample treatment can produce such effects [5, 6].

Generally, ultrasonication aids chemical analysis by:

- enhancing solid–liquid elemental extraction;
- shortening sequential extraction schemes for elemental determination;
- shortening elemental speciation schemes;
- speeding up solid–liquid extraction of organic species;
- accelerating electroanalytical measurements by enhancing mass transport efficiency;
- speeding up enzymatic reactions;
- accelerating liquid–liquid extraction techniques;
- enhancing the performance in solid-phase extraction and microextraction;
- incrementing accuracy in the solid-matrix dispersion technique.

1.2.1
Parameters Affecting Ultrasonic Cavitation

Ultrasonic cavitation is a physical phenomenon whose performance depends upon the parameters described below.

1.2.1.1 Frequency

At high sonic frequencies, on the order of the MHz, the production of cavitation bubbles becomes more difficult than at low sonic frequencies, of the order of the kHz. To achieve cavitation, as the sonic frequency increases, so the intensity of the applied sound must be increased, to ensure that the cohesive forces of the liquid media are overcome and voids are created. This phenomenon can be easily understood by looking at Figure 1.3, which shows the variation in threshold frequency versus intensity for aerated water and air-free water. As can be seen, ten times more power is required to induce cavitation in water at 400 kHz than at 10 kHz. The physical explanation for this lies in the fact that, at very high frequencies, the cycle of compression and decompression caused by the ultrasonic waves becomes so short that the molecules of the liquid can not be separated to form a void and, thus, cavitation is no longer obtained.

Recent literature suggests that ultrasonic frequencies can be of paramount importance for some analytical applications. For instance, Rial-Otero *et al.* have shown that

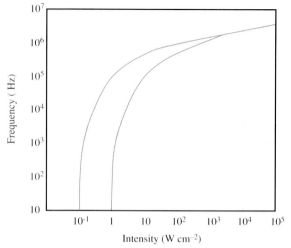

Figure 1.3 Variation of intensity of sonication versus the threshold frequency for aerated water (left-hand graph) and air-free water. Adapted from Ref. [1].

some pesticides can be determined in honey through a fast solid phase microextraction procedure using an ultrasonic bath operating at 130 kHz, whilst poor pesticides recoveries were obtained at an operating frequency of 35 kHz [7]. As another example, Pena-Farfal et al. have found that the frequency of the ultrasonic bath is critical for the enhancement of the enzymatic extraction of metals from mussel tissue [8].

1.2.1.2 Intensity

The intensity of sonication is proportional to the amplitude of vibration of the ultrasonic source and, as such, an increment in the amplitude of vibration will lead to an increase in the intensity of vibration and to an increase in the sonochemical effects. To achieve the cavitation threshold a minimum intensity is required. This means that higher amplitudes are not always necessary to obtain the desired results. In addition, high amplitudes of sonication can lead to rapid deterioration of the ultrasonic transducer, resulting in liquid agitation instead of cavitation and in poor transmission of the ultrasound through the liquid media. However, the amplitude should be increased when working with samples of high viscosity, such as blood. This is because as the viscosity of the sample increases so does the resistance of the sample to the movement of the ultrasonic device, for instance the tip of an ultrasonic probe. Therefore, a high intensity (i.e., high amplitude) is needed to set the ultrasonic device to obtain the necessary mechanical vibrations so as to promote cavitation in the sample.

Different authors have studied extensively the effect of sonication amplitude on the efficiency of the solid–liquid extraction of metals. There is agreement in the published data on this subject and, as a general rule, the higher the amplitude the more analyte is extracted, for given conditions of solution acidity and sonication time. Thus, the analyte extraction increases continuously with amplitude until an optimum is reached [9–11]. This effect of amplitude on the solid–liquid extraction of metals can be extended to the other ultrasonic applications. However, equilibrium should be attained between the intended effects and sonication amplitude since, as explained above, high amplitudes lead to high sonication intensities and high sonication intensities can promote some undesired effects, such as analyte degradation. Finally, the use of high amplitudes does not always lead to good results. For example, chromium is difficult to extract from biological samples using ultrasonication, whatever the amplitude used for an ultrasonic probe with a frequency of 20 kHz and a potency of 100 W [12].

1.2.1.3 Solvent
The solvent used to perform sample treatment with ultrasonication must be carefully chosen. As a general rule, most applications are performed in water. However, other less polar liquids, such as some organics, can be also used, depending on the intended purpose. Both solvent viscosity and surface tension are expected to inhibit cavitation. The higher the natural cohesive forces acting within a liquid (e.g., high viscosity and high surface tension) the more difficult it is to attain cavitation [2].

1.2.1.4 Temperature
Solvent temperature plays two roles in ultrasonication. On the one hand, the use of high temperatures helps to disrupt strong solute–matrix interactions, which involve Van der Waals forces, hydrogen bonding and dipole attractions between the solute molecules and active sites on the matrix. Moreover, faster diffusion rates occur at higher temperatures. On the other hand, cavitation is better attained at lower temperatures when the ultrasonic power of the generator is constant (Ref. [3], p. 68). This is because as the temperature of the solvent rises so to does its vapor pressure and so more solvent vapor fills the cavitation bubbles, which then tend to collapse less violently, that is, the sonication effects are less intense than expected. Hence a compromise between temperature and cavitation must be achieved. For example, the extraction ratios of polycyclic aromatic hydrocarbons from sediments were increased by between 6% and 14% when ultrasonic extraction with a probe was carried out under non-cooling conditions [13]. In contrast, when we tried to accelerate the enzymatic digestion of proteins in acetonitrile using an ultrasonic probe at room temperature, the enhancement in temperature caused by ultrasonication led to rapid evaporation of the solvent, making sample treatment impossible (J.L. Capelo *et al.*, unpublished data).

1.2.1.5 External Pressure and Bubbled Gas
If the external pressure is increased, then a greater ultrasonic energy is required to induce cavitation, that is, to break the solvent molecular forces. In addition, there is an increment in the intensity of the cavitational bubble collapse and, consequently, an enhancement in sonochemical effects is obtained. For a specific frequency there is a particular external pressure that will provide an optimum sonochemical reaction [2]. It must be stressed that most ultrasonic applications in analytical chemistry are performed under atmospheric pressure.

Dissolved gas bubbles in a fluid can act as nuclei for cavitation, favoring the ultrasonication process. However, ultrasonication can be used to degas a liquid. For this reason, if gas is used to increase cavitation effects it must be bubbled continuously into the solvent to maintain the effect. Monoatomic gases such as He, Ar and Ne should be used [1–3].

1.2.1.6 Direct and Indirect Ultrasonic Application

Ultrasonication can be applied in analytical chemistry in two ways: directly to the sample or indirectly through the walls of the sample container. Direct application is achieved through ultrasonic probes, which are immersed into sample, performing ultrasonication directly over the solution without any barrier to be crossed by the ultrasonication wave other than the solution itself. This approach has several drawbacks. For instance, sample contamination with metals detaching from the probe can be expected. Although modern ultrasonic probes are made from high purity titanium, contamination by metals such as Cr or Al has been reported [14]. Modern ultrasonic probes made from glass greatly reduce this problem [15]. Another disadvantage arises from the fact that most ultrasonic probes are used in open approaches, that is, the sample container is not sealed during sample treatment. Consequently, some volatile analytes can be lost. As an example, when the content of the 16 polycyclic aromatic hydrocarbons in the Environmental Protection Agency (USEPA) priority list was studied in sediments, a significant fraction of the most volatile compounds was lost due to the heating produced by the sonication probe used for the solid–liquid extraction process [13].

Indirect application is performed, generally, using an ultrasonication bath, although modern approaches take advantage of the powerful sonoreactor [15]. In both cases the ultrasonic wave needs first to cross the liquid inside the ultrasonic device and then to cross the wall of the sample container. Therefore, ultrasonication intensity inside the sample container is lower than expected. As ultrasonic baths are not powerful devices, their applications are greatly limited by the lack of ultrasonic intensity. In fact, many ultrasonic applications related to baths can be linked to the heating produced in the liquid that the bath contains – heat that is transmitted to the sample – rather than to actual ultrasonic effects, that is, cavitation [16]. However, as mentioned above, nowadays, a sonoreactor can be used instead of an ultrasonic bath in some regular applications [15]. A sonoreactor works like a powerful, small ultrasonic bath; it has been shown to be a reliable approach for speeding up chemical reactions in modern areas of analytical chemistry, such as protein identification and metal speciation [17, 18].

1.3
Common Ultrasonic Devices Used in Analytical Chemistry

1.3.1
Ultrasonic Bath

Figure 1.4 shows a modern ultrasonic bath equipped with the most advanced tools that are nowadays incorporated to improve performance, and which are commented on in the sections below.

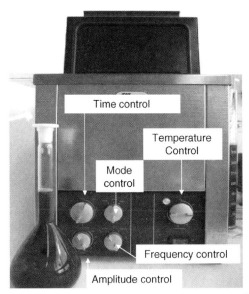

Figure 1.4 A modern ultrasonic bath incorporating the following advances: dual frequency of sonication, variable amplitude of sonication and three different modes of sonication.

1.3.1.1 Types of Ultrasonic Baths

At present there are three classes of ultrasonic baths [15]. The classic one is the common ultrasonic bath, which is found in most laboratories. This bath works with only one frequency, generally 40 kHz, and can be supplied with temperature control. A second type is provided as a multifrequency unit, which operates using, simultaneously, ultrasonic transducers with different frequencies, for instance 25 and 40 kHz, on the bottom and the side, respectively. The benefit of this is a uniform ultrasonic power distribution [19]. The third model corresponds to the most advanced in terms of technology, including the following features [20]:

1. Dual frequency of sonication.
2. A choice of 25/45 or 35/130 kHz. The baths are designed to work with one of the two frequencies at a time.
3. Power regulation.
4. The intensity of sonication can be controlled through amplitude control (10–100%).
5. Three operation modes:
   (a) Sweep: in this mode the frequency varies within a defined range. In this manner the ultrasonic efficiency is more homogeneously distributed in the bath than during standard operation.
   (b) Standard
   (c) Degas: the power is interrupted for a short period so that the bubbles are not retained by the ultrasonic forces.
6. Heating and timer.

1.3.1.2 How to Locate the Most Efficient Place Inside an Ultrasonic Bath

The ultrasonic intensity distribution inside an ultrasonic bath is not homogeneous. However, simple, rapid methods have been developed to locate the position that has the highest intensity of sonication (Ref. [3], p. 44). The aluminium foil test is, perhaps, the easiest method to apply in the laboratory. Using a series of aluminium foil sheets the most intense zones of sonication inside the bath can be quite accurately identified. As consequence of cavitation the aluminium foils are perforated (Figure 1.5). The maximum perforations occur at maximum intensity. Obviously, the reaction vessel should be located at the point where the maximum sonochemical effect is achieved.

1.3.1.3 Temperature Control

Most ultrasonic bath applications are performed for longer than 30 min and, as consequence of continuous ultrasonication, the bulk liquid warms up. Endothermic reactions can take advantage of this warming. In addition, the kinetics of many

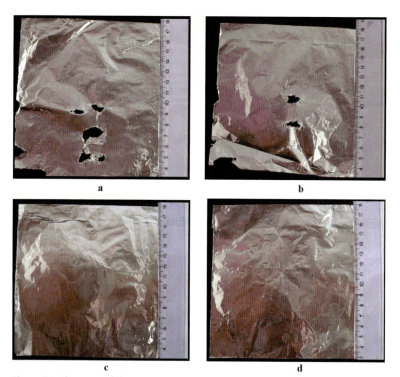

Figure 1.5 Aluminium foil test. Sonication time used for all experiments: 5 min. (a) 35 kHz sonication frequency and 100% sonication amplitude; (b) 35 kHz sonication frequency and 50% sonication amplitude; (c) 130 kHz sonication frequency and 100% sonication amplitude; and (d) 130 kHz sonication frequency and 50% sonication amplitude.

reactions are accelerated when the temperature is increased. However, some problems can arise regarding effectiveness and reproducibility of the sonochemical results obtained. Most ultrasonic baths are used without temperature control, and since bulk liquid warming is a slow process the final temperature achieved for a given time of sonication depends on the temperature of the laboratory (i.e., summer or winter). Thus, it is generally accepted that it is necessary to determine the maximum temperature the bath reaches and maintains, the so-called equilibrium temperature, when operating continuously under ambient conditions. Most reactions can be performed under the equilibrium temperature simply by filling the bath with water heated to that temperature previously. Alternatively, the problem can be solved by acquiring an ultrasonic bath with heater. However, before performing an ultrasonic treatment, it is recommended to wait until the working temperature is achieved.

When operation at room temperature is mandatory, control can be achieved with a simple water cooling recirculation system attached to the ultrasonic bath.

1.3.1.4 Shape and Material of Reaction Container

The shape of the reaction vessel is critical for the correct application of ultrasonication with a bath [2, 3]. This is because, as with any other wave, when the ultrasonic wave impinges against any solid surface some energy is reflected. If the base of the container is flat, such as in a conical flask, the ultrasound reflected is a minimum. Conversely, when the base of the container is spherical the ultrasonic wave hits the container at an angle, and a huge proportion of the ultrasonic wave is reflected away. For example, the amount of iodine liberated from KI in a solution of tetrachloride after 10 min ultrasonication, in a 100 mL flask, was three times higher when the flask had a flat base than when it had a round-bottom base [21].

The intensity of ultrasound is attenuated as it progress through a medium. The extent of attenuation is inversely related to the frequency. Therefore, the thickness of the wall of the vessel container should be kept to a minimum to avoid intense attenuation. This problem must be also borne in mind when the objective is to favor the solid–liquid extraction of analytes from solids deposited on a column: inside the column the effectiveness of the ultrasonic wave diminishes as it passes through the solid to the inner part of the column. Consequently, large-thin columns are preferred to short-wide ones for in-column ultrasonic bath applications.

1.3.2
The Ultrasonic Probe

Whilst the vessel container is immersed in an ultrasonic bath, an ultrasonic probe is immersed directly into the sample container. This is the main difference between the ultrasonic systems. The second difference is that the ultrasonic probe can deliver much higher ultrasonication intensity than the ultrasonic bath (100 times greater). These differences make each system appropriate for a different set of applications. The probe is generally used to attain effects that can not be achieved with the ultrasonic bath, for instance the mercury and arsenic speciation in human urine or seafood samples, respectively [22, 23].

1.3.2.1 Parts of an Ultrasonic Probe

Figure 1.6 shows the different parts of an ultrasonic probe. The generator converts mains voltage into high frequency electrical energy (generally, 20 kHz). The ultrasonic converter transforms electrical energy into mechanical vibrations of fixed frequency. The standard and booster horns increase the sonication amplitude. The probes or detachable horns transmit ultrasonic energy into the sample. The detachable horn design is crucial for a good performance. There are dedicated probes for a given range of volumes, and therefore mistakes such as using small probes for high volumes or high probes for small volumes must be avoided. The most important part of the whole system is the probe or detachable horn, which allows the vibration of the booster horn to be transmitted through a further length of metal in such a way that the power delivered is magnified.

Power magnification depends on the shape of the probe (Figure 1.7). The stepped probe gives the highest amplitude magnification [i.e., power, amplitude gain $(D/d)^2$)] of the shapes shown. Nevertheless, the exponential probe shape, although difficult to manufacture, offers small diameters at its working end, which makes it particularly suited to micro-applications.

1.3.2.2 Types of Ultrasonic Probes

Ultrasonic probes are usually made of titanium alloy (titanium probes) and are thermoresistant, can be treated in autoclaves and are resistant to corrosive media. The sample volume to be treated along with the sample type is crucial in determining

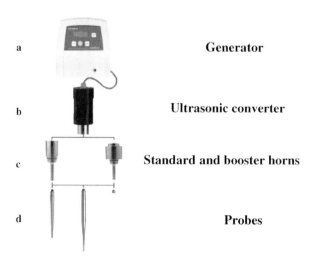

Figure 1.6 An ultrasonic probe. (a) Generator: the generator converts mains voltage into high frequency (20 kHz) electrical energy (most likely, although other frequencies are also available); (b) the converter transforms electrical energy into mechanical vibrations of fixed frequency, normally 20 kHz; (c) standard and booster horns: the horns increase the sonication amplitude; (d) probes: (also called detachable horns) probes transmit ultrasonic energy into the sample. The design is crucial for a good performance. Adapted from the Bandelin company with its kind permission.

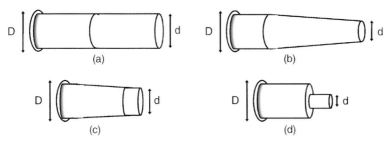

Figure 1.7 Probe shapes: (a) uniform cylinder; (b) exponential taper; (c) linear taper or cone; (d) stepped.

the selection of unit and the type of probe. It must be always borne in mind that the higher the amplitude provided by the probe the more intense is the sonication. Probes made of titanium alloy can contaminate the solutions in which they are used with metals such as Al or Cr [14]. For this reason, researchers working in trace metal analysis have long demanded the development of ultrasonic probes made from a non-contaminant material. Thus, the so-called silica ultrasonic probe made, as its name indicates, from silica has been developed [19]. This kind of probe has the advantage of non-particle metal intrusion in the sample. In addition, the typical metal contamination from titanium probes is avoided. Silica probes have high chemical and temperature shock resistance. Moreover, they do not exhibit electrical conductivity. However, several drawbacks must be addressed. For instance, the cavitation strength of silica glass is relatively low and, consequently, the amplitude must be limited. All silica glass probes are designed to be used with a maximum power setting of 50%, which corresponds to maximum amplitude of 12 µm. In addition, such probes are especially fragile during operation, and so they must not be placed on a solid surface or allowed to make contact with the vessel.

Spiral probes [19], made of Ti, Al and V, provide gentle ultrasonic treatment of aqueous media in test tubes or other long, thin laboratory containers. In contrast to traditional probes, the ultrasonic power is distributed across the entire surface of the spiral probe, thus making the distribution of the sonication intensity more homogeneous throughout the probe length. The drawbacks are related to metal contamination, like all probes made of titanium.

Multiple probes (Figure 1.1e and f) are manufactured to allow the use of two or more probes at the same time. The main advantage of this type of sonicator is throughput, which is greatly increased. Such sonicators meet the United States Environmental Protection Agency, USEPA, requirements specified in Method SW846-3550, which is referenced by the USEPA analytical test methods 8040, 8060, 8080, 8090, 8100, 8120, 8140, 8250, 8270, 413.2 and 418.1.

1.3.2.3 Dead Zones

The ultrasonic intensity rapidly decreases both radially and axially from the ultrasonic probe. For this reason the space between the ultrasonic probe and the wall of the container must be kept to a minimum, while ensuring that the probe does not touch the container – otherwise the probe might break. Keeping dead zones to a minimum

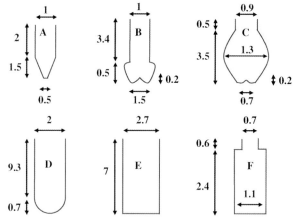

Figure 1.8 The shape of the vessel influences ultrasonic performance. The best forms are those that minimize "dead zones," namely, forms A–C. Adapted from Ref. [13].

ensures a maximum contact between the sample and the cavitation zones, and also among the sample particles, which helps to diminish their size by collisions and hence increases the total area in contact with the solution. The final result will be an increment in desired effects, such as, for example, in the solid–liquid extraction efficiency. Figure 1.8 shows different types of vessels. The influence of their shape on the extraction of PAHs from sediments under an ultrasonic field has been studied. The distance between the probe and the wall container in the case of design D is 8.5 mm and for desing E it is 12 mm, whereas the distances for containers A, B, C and F are 3.5, 3.5, 3 and 2 mm, respectively. Sample treatments carried out with containers D and E led to recoveries below 70%. This low extraction efficiency may be linked to the "dead cavitation zones", due to the distance between the probe and the wall container.

1.3.2.4 How to Choose the Correct Ultrasonic Probe

Two main issues determine the type of ultrasonic probe chosen for laboratory work. The type of application is the primary concern. As an example, if we are going to work with elemental extraction from soft tissues, then the probe has only one requirement: to be made of silica to avoid metal contamination. If the metal contamination introduced in solution by titanium probes is not a problem, and the elemental extraction is to be carried out in samples with low solid–liquid elemental extractability under the effects of an ultrasonic field, such as sediments or soils, then the most powerful ultrasonic probe made from titanium should be chosen. The high intensity will ensure both stronger and a higher number of collisions among particles, thus increasing the total area of contact between the solid and the solvent, and raising the ratio of elemental extraction. The second concern is related to the sample volume to be sonicated. The lowest volume that can be sonicated with an ultrasonic probe is 10 µL, but to date only the Dr Hielscher company offers a dedicated ultrasonic probe for such a volume [24].

1.3.2.5 Temperature Control

The utilization of an ultrasonic probe leads to an increase in bulk temperature. The higher the amplitude the faster the temperature increases. If the temperature is not controlled some undesired effects can occur. The most obvious is the degradation of compounds of interest, but also the volatilization of low volatile analytes can occur [13]. In addition, as the temperature is increased, the physical characteristics of the liquid media change in such a way that the ultrasonic transmission can be affected and no cavitation is achieved. This phenomenon is known as "decoupling" [2, 3]. It is not easy to control the temperature when ultrasonication is applied with an ultrasonic probe. However, three strategies can be followed. The first and simplest consists in using an ice bath. The vessel is inserted in an ice bath and sonication is performed. This ensures a rapid dissipation of heating. This strategy can be applied for short sonication times, otherwise care is needed to replace the ice. The second strategy entails the use of dedicated vessels, such as those shown in Figure 1.9, which are specially designed to dissipate warming. The third strategy is to use the "pulse" mode of ultrasonic application, which is available in all modern

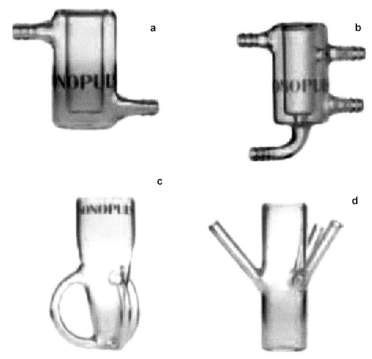

Figure 1.9 Vessels specially designed for applications with an ultrasonic probe. The design allows rapid refrigeration (Reproduced with the kind permission of Bandelin). (a) Cooling vessel KG for sonication of temperature-sensitive samples; (b) flow-through vessel with cooling jacket for irradiation of large volumes; (c) Rosett cell RZ for homogenous and intense recirculation of sample; (d) Suslick cell SZ with three sidearms for the introduction of gas or measuring probes.

sonication probes. In this mode the amplifier switches the power of the probe on and off repeatedly, thus avoiding the build up of reaction temperature. The three strategies can be combined.

1.3.2.6 Shape and Material of Reaction Container

A critical factor to be considered, for a correct ultrasonic probe application, is the shape of the reaction vessel, which must be conical [7] and with a diameter as small as possible so as to raise the liquid level for a given volume. When the concept of analytical minimalism is applied to any analysis, the low sample amount and solution volume used require a dedicated shaped vessel. The influence of vessel shape on the extraction of PAHs (Figure 1.8) from sediments using an ultrasonic probe has been studied [13]. Design A in Figure 1.8 corresponds to a classical Eppendorf cup made from polypropylene, while designs B and C each have a bump in the base and were laboratory-made from glass. These latter designs assist sonic mixing by dispersing the sonic waves as they impinge upon the bump and are reflected from the base. Design D has a spherical bottom whilst designs E and F each have a flat-bottom. Receptacles A, B, C and F were specially designed to work with low volumes, to reduce the dead zones during sonication. Moreover, the distance from the probe sonicator to the wall container was minimized to maintain the particles inside the area under the effects of cavitation. Cavitation phenomena rapidly decrease axially and radially. The low sonication volume allows to induce more collisions among particles and, hence, increase the total area exposed to solvent extraction. Receptacle C was used with a volume of 4 mL, whereas D was used with larger volumes (5 mL and 18 mL). Vessels A, B and C were found to be appropriate to achieve total extractions, whereas the extraction efficiency using design F was the lowest of all attempts made with containers with a total volume lower than 2 mL. In the latter case, due to the flat bottom, the particles climbed up the ultrasonic tip, leading to poor particle–solution contact and, hence, to lower extraction efficiencies.

Regarding the material of the reaction container, the particular analytical application will dictate the best one. Thus, if the analyte under investigation is a phthalate, then plastic containers should be avoided.

1.4
Current Ultrasonic Devices for New Analytical Applications

Cup horns (Figure 1.1h) are available from almost all ultrasonic companies. The sonoreactor (Figure 1.1d) is available from the Dr Hielscher Company. Such ultrasonic devices offer indirect sonication. This means that the ultrasonic waves need to cross the wall of the sample container. This does not occur with the ultrasonic probe, which is immersed directly into the sample, giving direct sonication. Cup horns, the sonoreactor and microplate horns can be compared to high intensity ultrasonic water baths. For example, the sonoreactor is 50 times more intense for a 2 mL volume of sonication than an ultrasonic bath. Using the above-mentioned devices, samples can be processed in sealed tubes or vials, eliminating aerosols, cross-contamination or the

Table 1.1 Important ultrasonic companies along with the main parameters of the most common ultrasonic devices.

Variable	Sonoreactor	Ultrasonic probe	Ultrasonic bath
Sonication time (s)	Up to 300	Up to 120	Up to several hours
Thermostat	No	No	Yes
Intensity of sonication at 1.5 mL vial (W)	0.5	15	0.01
Amplitude (%)	20–100	10–100	10–100
Sample handling	Low	High	Low
Sample throughput (1 mL vials at once)	Up to 6	Up to 96 (multiple probes)	100 to 500 (depending on size)
Direct application	No	Yes	No
Solid–liquid extraction yield	Medium	High	Low
On-line applications	Yes	Yes	Yes
Degradation of organics	Medium	High	Low
Ultrasonic companies	www.bandelin.com; www.hielscher.com; www.equilabcanada.com; www.bransonultrasonics.com; www.misonix.com; www.elmaultrasonic.com		

loss of volatile compounds. Such devices are ideal for samples such as those that are sterile or dangerously pathogenic. In the cup horn, the titanium probe is held within an acrylic cup filled with water. Samples are placed within the cup, above the probe. The cavitation produced in the immersed samples is higher than that given by an ultrasonic bath but it is lower than the cavitation produced by direct immersion of the ultrasonic probe into the solution. All the systems allow refrigeration. The microplate horn is similar to the cup horn, but it allows high-throughput applications, especially for the 96-well plate. Different versions of the same concept can be found for several of the ultrasonic companies cited in Table 1.1.

References

1 Mason, T.J. and Lorimer, J.P. (1989) *Sonochemistry: Theory, Applications and uses of Ultrasound in Chemistry*, Wiley-Interscience, New York.

2 Mason, T.J. (1992) *Practical Sonochemistry: User's Guide to Applications in Chemistry and Chemical Engineering*, Ellis Horwood Ltd, New York.

3 Mason, T.J. (2000) *Sonochemistry*, Oxford Chemistry Primers, Oxford, UK.

4 Suslick, K.S., Cline, R.E. and Hammerton, D.A. (1986) *Journal of the American Chemical Society*, **108**, 5641.

5 Wibetoe, G., Takuwa, D.T., Lund, W.D. and Sawula, G. (1999) *Fresenius' Journal of Analytical Chemistry*, **363**, 46.

6 Capelo-Martınez, J.L., Ximenez-Embun, P., Madrid, Y. and Camara, C. (2004) *Trends in Analytical Chemistry*, **23**, 331.

7 Rial-Otero, R., Gaspar, E.M., Moura, I. and Capelo, J.L. (2007) *Talanta*, **71**, 1906.

8 Pena-Farfal, C., Moreda-Pineiro, A., Bermejo-Barrera, A. et al. (2004) *Analytical Chemistry*, **549**, 3541–3547.

9 Amoedo, L., Capelo, J.L., Lavilla, I. and Bendicho, C. (1999) *Journal of*

Analytical Atomic Spectrometry, **14**, 1221–1226.
10 Lima, E.C., Barbosa, F., Krug, F.J. *et al.* (2000) *Journal of Analytical Atomic Spectrometry*, **15**, 995–1000.
11 Capelo, J.L., Lavilla, I. and Bendicho, C. (1998) *Journal of Analytical Atomic Spectrometry*, **13**, 1285–1290.
12 Amoedo, L., Capelo, J.L., Lavilla, I. and Bendicho, C. (1999) *Journal of Analytical Atomic Spectrometry*, **14**, 1221–1226.
13 Capelo, J.L., Galesio, M.M., Felisberto, G.M. *et al.* (2005) *Talanta*, **66**, 1272–1280.
14 Wibetoe, G., Takuwa, D.T., Lund, W. and Sawula, G. (1999) *Fresenius' Journal of Analytical Chemistry*, **363**, 46–54.
15 Santos, H.M. and Capelo, J.L. (2007) *Talanta*, **73**, 795–802.
16 Patrício, A., Fernández, C., Mota, A.M. and Capelo, J.L. (2006) *Talanta*, **69**, 769–775.
17 Rial-Otero, R., Carreira, R.J., Cordeiro, F.M. *et al.* (2007) *Journal of Proteome Research*, **6**, 909.
18 Vale, G., Rial-Otero, R., Mota, A. *et al.* (2008), *Talanta*, **75**, 872–884.
19 BANDELIN electronic, GmbH & Co. KG, Heinrichstraße 3-4, D-12207, Germany BerlinTwinSonic® series, www.bandelin.com, last accessed 7 November 2007.
20 Elma Hans Schmidbauer GmbH & Co, KG Postfach 280 D-78202. Singen Kolpingstr. 1-7, D-78224 Singen, Germany; www.elma-ultrasonic.com, last accessed 21 November 2007.
21 Mason, T.J., Lorimer, J.P., Cuesta, F. and Paniwnyk, L. (1989) Ultrasonics International 89, Conference Proceeedings, p. 1253.
22 Capelo, J.L., Lavilla, I. and Bendicho, C. (2000) *Analytical Chemistry*, **72**, 4979.
23 Capelo, J.L., Lavilla, I. and Bendicho, C. (2001) *Analytical Chemistry*, **73**, 3732.
24 Hielscher Ultrasonics GmbH, Am Dobelbach, 19 D-70184 Stuttgart, Germany; www.hielscher.com, last accessed 21 November 2007.

2
Ultrasonic Energy as a Tool for Sample Treatment for the Analysis of Elements and Elemental Speciation

Hugo Miguel Santos, Carlos Lodeiro, and José-Luis Capelo-Martínez

2.1
Introduction

This chapter is devoted to one of the most important applications of ultrasonic, US, energy in the analytical laboratory: the analysis of elements. The analysis of elements entails different approaches that are linked to the knowledge sought for a given sample. Thus, the total content of an element, for instance total As, or the chemical forms in which the element is present in the sample, for instance arsenobetaine, is information that can be obtained with common analytical techniques such as atomic absorption spectrometry following element solubilization. Each required analysis, total content or chemical speciation, involves dedicated sample handling. This handling can in most cases be simplified using ultrasonic energy. The most important parameters and applications of ultrasonic energy for sample handling for total element content and elemental speciation are treated in detail.

2.2
Parameters Influencing Element Ultrasonic Solid–Liquid Extraction

2.2.1
Extracting Reagent

The choice of adequate extracting reagent(s) along with the correct concentration(s) is critical for an adequate performance when developing an element extraction protocol using ultrasonication [1]. As a general rule, a lower reagent concentration is always desirable for both safety and environmental reasons. However, the latter principle does not always match high extraction efficiency. Overall, element extraction can be classified into three categories, as a function of the questions to be answered regarding metal content: (i) total content, (ii) metal speciation and (iii) sequential extraction schemes . Table 2.1 shows the most commonly used extracting reagents

Ultrasound in Chemistry: Analytical Applications. Edited by José-Luis Capelo-Martínez
Copyright © 2009 WILEY-VCH Verlag GmbH & Co. KGaA, Weinheim
ISBN: 978-3-527-31934-3

Table 2.1 Type of intended extraction and most common reagents used for each one.

Type of extraction	Reagents used	Comments
Total Extraction	HNO_3, HCl, H_2O_2	Combinations of these reagents are also used. Other reagents reported are H_2SO_4; $HClO_4$; and HF (*Caution*: HF is an extremely dangerous chemical and must be handle with care)
Speciation	HCl, HClO, enzymes, and organic chemicals, such as cyclohexane	As a general trend, acids are used in low concentrations
Sequential extraction	Acetic acid, H_2O_2, ammonium acetate, HF, HNO_3	Reagents chosen are a function of SE schemes. Low amplitudes and low sonication times are needed

and reagent combinations for these categories in which metal extraction studies were classified.

2.2.1.1 Extracting Reagents for Total Element Extraction

Total element extraction using ultrasonication is achievable for soft biological tissues and for some inorganic matrixes. However, not all elements can be extracted even for these matrixes and so, if possible, the use of ultrasonication in evaluating total element content should be carefully compared with available literature data, concerning both the element and matrix of interest. If no relevant literature information is available, then another analytical sample treatment should also be used for comparison to ensure adequate performance of the ultrasonic method. For instance, the matrix can be dissolved with the microwave pressurized acid digestion, MWPAD, procedure and the element content compared with that obtained by ultrasonic extraction [2, 3].

Extracting agents generally used in US extractions for total content include acids, such as HNO_3 or HCl, and oxidizing agents, such as H_2O_2. These reagents can be used alone or in combination. Although other chemicals can be used, the ones mentioned above are, by far, the most cited in literature [1, 4, 5].

Selecting the correct reagent combination is critical if accurate results are demanded. As an example, Ashley *et al.* have found different extraction efficiencies from the same reference environmental samples as a function of the extracting reagent as follows: for Ba, from 35 to 101% and from 20 to 56% when HNO_3 (25% v/v) and HNO_3 : HCl (1 : 1 v/v) were used, respectively; and for Co from 52 to 58% and from 68 to 100% for the same extracting reagents [6].

The reagent concentrations vary as a function of the ultrasonic device used. As a general rule, for the same element and matrix, an ultrasonic bath (UB) requires higher concentrations than an ultrasonic probe (UP) to obtain comparable extraction efficiencies. Typical acid concentrations for probe sonication are less than 1 M while for UBs acid concentrations as high as 50% v/v have been reported [3].

A feature to be considered in the literature is analyte re-adsorption as a function of the acid concentration used for the ultrasonic extraction. Thus, Se has been extracted (100%) from seafood samples with an UP and HNO_3 (0.5% v/v) as extracting reagent. Interestingly, for higher HNO_3 concentrations less Se was extracted. The authors hypothesized that this is probably due to re-adsorption rather than to a lack of ultrasonic extractability [7].

2.2.1.2 Extracting Reagents for Elemental Speciation

As we will see in Section 2.8, chemical speciation for elements is still under development. The reagents used for elemental speciation in conjunction with ultrasonication can be classified into two categories: inorganic and organic.

The inorganic category includes acids or acid mixtures at low concentrations. This type of speciation is scarce, since it is difficult to extract element species while preserving their integrity using acids and ultrasonication [5]. However, using this approach it is possible to distinguish between inorganic and organic species or between toxic and less toxic species. Thus, using HCl and HClO, inorganic mercury and organic mercury can be differentiated in drinking water, seawaters and wastewaters [8, 9] by flow injection cold vapor atomic absorption spectrometry, FI-CV-AAS. As another example, using O_3 and HCl, it is possible to determine reactive arsenic towards sodium tetrahydroborate by FI-hydride generation HG-AAS [10], making it possible to distinguish between toxic arsenic forms, such as As(III) and As(V), and less toxic As species, such as arsenobetaine; the latter is the most abundant As species in fish and shellfish.

The organic category includes the use of organic solvents to separate the species to be determined from the matrixes of interest. This category also includes the use of enzymes to separate element species from biological samples such as animal and botanical tissues. The latter type of speciation has been improved greatly by combining ultrasonication and enzymes in a relatively new, promising sample treatment called ultrasonic assisted enzymatic digestion, USAED [11, 12]. Before employing USAED, however, it must be stressed that some factors critical to attaining reliable results need to be carefully considered. For example, the correlation between the samples and the enzymes, the pH of the solution, the temperature of the bulk solution, the ratio of substrate to enzyme, the enzyme ageing and cleaning procedures need to be carefully optimized along with the common variables for sample treatments making use of ultrasonication, such as type of sonication device, sonication time, sonication amplitude, sonication frequency and type of container [12].

2.2.1.3 Extracting Reagents for Sequential Extraction Schemes

Nowadays it is well recognized that the toxicity of a pollutant is directly linked to its environmental mobility, which is dictated by its chemical form and binding state. Changes in environmental conditions, such as acidification or redox potential, can cause trace-element mobilization through the environment, that is, contamination of waters by toxic metals because of element mobilization through acid leaching of soils caused by acid rain. Therefore, identification of the main binding sites and phase associations of trace elements in soils and sediments helps in understanding the environmental mobility of elements [13]. To attain such information a technique known as fractionation or sequential extraction scheme (SES) is used. SES is based on the application of sequential selective chemical extractions using a series of more or less selective reagents, with the aim of simulating the various possible natural and anthropogenic modifications of the environment. The most common SESs are Tessier SES and the Standard, Measurement and Testing programme, SM&T, SES, from the Commission of the European Communities [14, 15].

Treatment time is the major drawback in SES. In fact, the Tessier and SM&T SESs take two to three days to complete. One of the advantages of ultrasonication is its capability to speed up the sequential steps of the SES procedures. The reagents used and their concentrations are exactly the same as in conventional SES. For instance, good agreement has been observed between the extractable amounts of Pb, Cd, Cu and Cr in sediments obtained with the SM&T SES standard procedure and those obtained with a small-scale ultrasound-assisted extraction procedure; the total time being reduced from 51 to 2 h [16].

2.2.2
Matrix Properties

2.2.2.1 Type of Matrix
The type of matrix along with the type of analysis (total extraction, speciation, SES) will dictate the extracting solution and the reagent concentrations (Figure 2.1) [1, 4]. Synthetic solids, such as resins used in metal pre-concentration procedures, and soft tissues, such as mussel tissue, are the type of matrixes from which element extraction in high percentages is easily achieved with low reagent(s) concentration(s), low sonication time and low sonication amplitude. Ultrasonic extractions in other biological tissues such as roots and needles are more metal-binding dependant; the extraction is achievable only for some metals, such as Cd, and under conditions of medium to high reagent(s) concentration(s) and high sonication intensities. For inorganic samples, extraction of metals by means of ultrasonication from sewage sludge perform very well, affording good recoveries, generally >80% for most metals. In contrast, soils and sediments are matrixes for which the ultrasonic extraction of metals is of doubtful utility since most literature data suggest low extraction rates for most elements [1, 4, 5].

2.2.2.2 Mass of Matrix
One of the advantages of ultrasonication is that it can be performed in such a way that minimum amounts of sample are needed, providing that the final element concentration lies in the linear range of the detection method [16, 17]. For a given volume, for instance 1 mL, 10–100 mg can be used, when ensuring that longer sonication times are used for higher amounts of sample. Importantly, for a constant volume, the higher the mass of the matrix used, the higher the concentration of substances co-extracted, for instance organic matter. A high content of organic

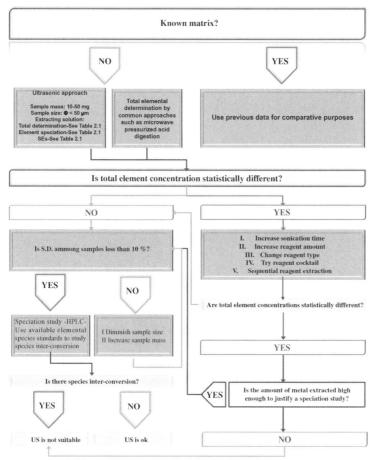

Figure 2.1 Comprehensive scheme for the development of analytical methodologies for element determination (total, speciation, SES) using ultrasonication.

matter in solution is a drawback for the correct performance of some analytical techniques, such as FI-CV-AAS or electrothermal-ET-AAS. If analytical minimalism is not an important issue for the analyst, sample amounts as large as 1–2 g in volumes such as 40 mL can be submitted to ultrasonic treatment. As an example, the same Pb concentrations were obtained after ultrasonic extraction with probe from tea leaves, aquatic plant and mussel tissue with sample/volume quantities of 0.01 g/1.5 mL, 0.1 g/15 mL and 0.250 g/40 mL [18]. It must be pointed out that different volumes of sonication require probes of different sizes, as explained in Chapter 1.

The homogeneity of the sample and the distribution of the analyte in the sample will also affect the mass of matrix used for the analysis. However, these two variables can be overcome by diminishing the sample size, as explained in the next section.

2.2.2.3 Sample Size

It is now well recognized that the particle size affects analytical performance in the ultrasonic solid liquid extraction of elements, US-SLE, of elements. If the analyte is not homogeneously distributed in the matrix, grinding samples to a small particle size (<de 30 μm) will increase the homogeneity of the suspension, thereby increasing the likelihood that a representative mean concentration can be obtained. In addition, as the particle size decreases, metal extraction is facilitated by increasing the contact area between the sample and the solvent. A direct relation between total area of matrix and analyte extraction can be found. To simplify the problem, it is assumed that matrix particles are spherical. Then, the percentage of metal extracted, E, for a given particle radius, r, is linked through a constant K as follows:

$$E = \pi K r^2$$

Once the constant has been established for a given matrix, it is possible to estimate the minimum particle size needed to achieve total analyte extraction. If the size required is too small, then the ultrasonic approach must be substituted by another sample treatment, such as MWPAD. As a practical example, about 70% of Cd was extracted from sediment with a particle diameter below 90 μm [19]. If we assume an average particle radius of 45 μm, K is 11.8×10^4. By using this constant, the minimum size required to obtain total extraction is less than 38 μm. This could explain not only why total recovery was not possible for this sample but also the fact that Lima et al. reported total Cd extraction from sediments with a particle size less than 30 μm under similar experimental conditions [20].

2.2.3
Ultrasonic Device

As recently pointed out, the technology for ultrasonication has evolved impressively since the beginning of the twenty-first century [21]. Nowadays, the most common ultrasonic devices used for analytical sample treatment, the UB, the UP and the sonoreactor (SR), are provided with the selection of ultrasonic amplitude, different ultrasonic frequencies can be chosen and the sonication time is also programmable. The characteristics of these ultrasonic devices are treated in detail in Chapter 1. As a general rule, high extraction efficiencies are achieved with the UP and the SR, as shown in the following sections.

2.2.3.1 Type of Ultrasonic Device

The UB, UP and the SR are the most frequently used systems in the analytical laboratory. Their characteristics along with their differences are summarized in Table 1.1. The following sections give some advice regarding the most important variables influencing US-SLE: time of ultrasonication, amplitude of sonication and frequency of ultrasonication.

2.2.3.2 Time of Ultrasonication

As a general rule, the time needed to perform US-SLE analytical applications varies with the intensity of sonication that an US device can provide. Thus, times as long as 30 min or longer are used in sample treatments making use of an UB, which is a low sonication intensity device. Times of between 5 and 10 min are regularly used with the SR, while times of 1 to 3 min are common for applications performed with the UP. The longer the time needed for an application, the quicker the sonication device is degraded. For long applications times, low amplitudes are recommended, to avoid rapid deterioration of the ultrasonic transducers.

2.2.3.3 Ultrasonic Amplitude

Generally, the ultrasonic amplitude chosen depends on the ultrasonic intensity to be delivered into solution. For a constant time, as the ultrasonic amplitude increases so do the cavitation effects and the intensity of sonication. As an example, for the same ultrasonication time, Cd was extracted from soils (100%) at amplitudes of 20% whilst Pb could only be extracted from the same soils at an amplitude of 40%, an increase in sonication power of 20% [18, 19]. In most cases, amplitudes of around 50% perform well.

2.2.3.4 Ultrasonic Frequency

Most modern ultrasonic devices used in analytical chemistry work at frequencies of between 20 to 40 kHz. Some UBs, however, can be used at two different frequencies, of 35 and 130 kHz, as explained in Chapter 1. Bearing in mind that higher intensities of sonication are obtained at low frequencies, frequencies of around 30 kHz are recommended for total element extractions. Some recent work has highlighted the importance of choosing high sonication frequencies to perform extractions intended for elemental speciation when using UBs, especially extractions in which enzymes are used to degrade organic matter and to extract the different chemical forms of the elements [22]. The importance of sonication frequency in elemental speciation is still under investigation and conclusive statements can not be made yet.

2.2.3.5 Temperature of Sonication

Theoretically, as the bulk solution temperature rises, the performance of the ultrasonic treatment deteriorates. This is because a lower cavitation efficiency is obtained as the density of the liquid decreases with the increasing temperature caused by ultrasonication. This phenomenon is explained in detail in Chapter 1. However, for some applications, non-cooling conditions can be desirable, since effects other than cavitation must also be considered. For example, when US-SLE of Se using enzymes was accelerated with an UP, cooling of the bulk solution resulted in recoveries that were at least a 20% lower than for extractions performed under non-cooling conditions [23]. This can be attributed to the fact that some enzymes perform better at temperatures in the range 35–60 °C than at room temperature. As a general rule, most ultrasonic extractions are carried out under non-cooling conditions.

2.3
US-SLE from Soils and Sediments

Soils and sediments are the most complicated matrices in terms of extraction of elements assisted by ultrasonication. The literature data is inconclusive, with the reported percentages of extraction varying greatly for a given element. Some examples are given in Table 2.2, where it can be seen that, for instance, Cr is only partially extracted from soils and sediments [6]. Other examples report 30% Cr extraction from sediment and extractions of 24–47% from sediments and soils [24, 25]. Although for some metals high recoveries from soils and sediments can be achieved, long sonication times, high sonication amplitudes and high reagents concentrations are needed. Even under such conditions, however, metals such as Co, Ni, Mn, Ba, Zn, V, Fe and Cu have been extracted in percentages below 75% from some soils and sediments [20, 26, 27].

2.4
US-SLE from Sewage Sludge

The recycling of sewage sludge for agricultural use is a useful environmental-friendly alternative to incineration, since the sludges act as a source of organic matter and inorganic nutrients. However, their utilization requires control of any organic and inorganic pollutants they may contain. Sewage sludge is a matrix with an important inorganic fraction. The higher the inorganic composition the more difficult it is to extract metals. Metals such as Cd, Co, Cr, Cu, Mn, Ni, Pb and Zn have been extracted successfully with ultrasonic energy from sewage sludge from wastewater treatment plants, using UP with a glass sonotrode (80% sonication amplitude, 20 kHz sonication frequency, 500 W, 20 min sonication time, $HNO_3:HCl$ 1:1 v/v extracting solution) [28] or UB (150 W, 10 min) [29]. Other authors, however, have found low recoveries for Cd (56 ± 1%) and Cr (47 ± 1%) in sewage sludge, probably due to a lack in extractability caused by the low acid concentration used in the extraction process (5% v/v HNO_3) [18]. Therefore, special care must be taken regarding the reagents used and their concentrations if total element extraction is the desired result.

2.5
US-SLE Extraction from Plants

The elements Ca, Mg, Mn, Zn, Co, Ni, Cu, Ba, Fe, Pb and Cd have been extracted from plants using either an UB or an UP with different degrees of success [30–36]. As a general trend, Ca, Mn, Mg and Cu are totally extracted. Fe, Cr, Zn and Ni are highly extractable, with recoveries ranging from 70 to 100%. Barium is extracted in low percentages [31]. The reagent used most often for US-SLE is HNO_3, with reported concentrations ranging from 0.05 to 5% v/v. It must be stressed that low acid concentrations, such as 0.05% v/v, are not recommended since the pH is critical in

Table 2.2 Elemental recoveries obtained after US-SLE from soils and sediments.

Reference	Matrix	Liquid medium Ultrasonic device[a]	Technique of analysis; analyte (% extracted); comments
[20]	Sediment	0.5% v/v HNO_3	ET-AAS; Cd (89–105); Cu (90–101); Pb (88–98). Particle size < 30 μm
[33][b]	Sediment	UP 5% v/v HNO_3 Probe, 100 W	ET-AAS; Cu (60); Cr (10)
[107][b]	SRM 2704 sediment	5% v/v HNO_3	ET-AAS; As (68); Fe (22); Mn (64); Pb (88)
[108]	Sediment	UP, 15 s 30% v/v HCl	HG-AAS; As (113); 15 min of centrifugation (7500 rpm) followed by filtration (0.45 μm). Particle size <50 μm
[109]	SRM 2710 Montana soil	Probe, 100 W, 20 kHz, 3 min, 80% 50% v/v aqua regia/water; UB; 650 W; 35 kHz; 9 min; 50 °C	ICP: As (99); Cu (98); Pb (93); Sb (59); Zn (85)
[26]	Soils and sediments	0.5% HNO_3 UP, 100 W	SS-ET-AAS; soils: Cr (75–88); Ni (82–95); V (40–102). sediments: Cr (46–59); Ni (68–86); V (91–100)
[110]	Four certified sediments	Aqua regia UB; 40 kHz; 30 min	SS-CVG-ICP; As (100–104); Hg (95–104); Sb (95–104); Se (99–117); Sn (91–98); particle size <120 μm
[111]	Contaminated soil	Citrate buffer UP 210 W, 19.5 kHz, 30 min	ICP; Pb (90); Zn (70.6); Cd (100); Cu (85)

(Continued)

Table 2.2 (Continued)

Reference	Matrix	Liquid medium Ultrasonic device[a]	Technique of analysis; analyte (% extracted); comments
[6]	Sediments and soils	(1) 25% v/v HNO$_3$; (2) HNO$_3$/HCl (1:1 v/v); (3) 25% HNO$_3$/HCl (1:1 v/v) UB; 1 W cm^{-2}; 60 min	ICP; (1) 25% v/v HNO$_3$: As (84–99); Ba (35–101); Cd (79–100); Co (52–58); Cr (13–105); Cu (85–94); Fe (31–104); Mg (62–109); Mn (71–107); Ni (49–107); Pb (88–108); Se (68–135); V (35–105); Zn (63–102). (2) HNO$_3$/HCl (1:1 v/v): As (76–130); Ba (20–56); Cd (87–122); Co (68–100); Cr (17–60); Cu (92–100); Fe (52–68); Mg (41–68); Mn (72–78); Ni (65–90); Pb (85–95); V (68–135); (3) 25% HNO$_3$/HCl (1:1 v/v): As (83–90); Ba (17–39); Cd (84–95); Co (48–58); Cr (14–43); Cu (80–94); Fe (30–33); Mg (24–64); Mn (63–70); Ni (39–52); Pb (91–95); V (13–62)

[a] When given, nominal sonication power, sonication time, sonication amplitude, sonication frequency, temperature of sonication.
[b] Data given as analyte partitioning, the sample being analyzed as a slurry.

obtaining accurate results. Thus, Zn recovery was raised from 23 to 74% from tea leaves when the HNO_3 concentration of the extracting solvent was increased from 0.04 to 4% v/v [31]; and Cd recovery was raised from 65 to 86% from cabbage leaves when the HNO_3 concentration was increased from 0.05 to 5% v/v. Some researchers have used other types of solvent mixtures to perform extractions from plants, such as HCl 1 M-Triton X-100 0.05% (w/v) [36]. An interesting strategy to increase the percentage of metal extracted consists of the addition of a chelating agent in an appropriate medium. The chelating agent acts as an in-solution trapping agent, favoring metal release from the solids. For instance, Filgueiras et al. have reported the successful US-SLE of trace and minor metals (Ca, Cd, Mg, Mn, Pb and Zn) from plant tissue using ethylenediaminetetraacetic acid in alkaline medium [37]. For some metals, some highly variable results can be found in the literature. For example, Pb recoveries from tea leaves, cabbage leaves and cabbage roots have been reported as 2, 1.5, and 1.5% respectively [31, 32]. These low extractions rates can be attributed to the low sonication time used, only 30 s, since other authors have reported total Pb recovery from beech leaves (reference material, RM, CRM 100) and from olive leaves (RM, CRM 062) with a sonication time of 300 s [36]. Another interesting finding of a literature survey is that US-SLE from metals from plant tissues seems to be independent of the ultrasonic device used, provided an adequate solvent and sonication time is used – longer sonication times are needed when extractions are done with an UB than with an UP.

2.6
Extraction from Soft Tissues

By soft tissues we mean most biological tissues, such as plankton, oyster or algae tissues [2, 39], and also mussel tissue [18], dogfish muscle and dogfish liver [19], lobster hepatopancreas [10], bovine liver [34], seafood samples [40], or breast cancer biopsies [41]. US-SLE of elements from this type of matrix is the recommended sample treatment for element determination. The metal bindings to these organic tissues are so weak that extraction is easily achievable for most elements in a 5% v/v HNO_3 solution, using UB or UP. In fact, short sonication times are needed to achieve total element extraction: about 5 min with an UP and between 10 and 30 min for an UB. The metals Cd [2, 19, 40], Pb [2, 18, 40], Mn [32, 40], As [10, 40], Se [39, 40] and Co, Cr, Cu, Mg and Zn [41] have been totally extracted in soft tissue samples with the aid of ultrasonic energy.

2.7
Total Element Determination

2.7.1
US-SLE and US-SS for F-AAS

The best way to join US-SLE and flame, F,-AAS is by first extracting the target elements and then separating the solid fraction from the liquid one, by centrifugation and, if necessary, filtration. With this simple sample treatment, US-SLE and F-AAS have been used by different authors to measure metals such as Ca, Fe, Mg, Mn and Zn [36, 37]. Potentially, all extractable metals can be measured by this approach, depending, of course, on their final concentration in solution.

F-AAS is a technique that must be handled with care since there is always risk of explosion. Solid powders can be introduced as suspensions of finely ground powder to measure the elements by F-AAS by using the slurry technique and ultrasonication, named ultrasonic slurry sampling, US-SS [40]. However, this procedure is not recommended unless the skills of the operator permit it as the introduction of solids in F-AAS is linked with the risk of obstruction of the nebulizer [42]. In addition, great care must be taken to keep the burner unblocked, and so special cleaning procedures must be adopted. Furthermore, both the sample uptake tube and the drain outlet can also be blocked if the sample is too big or the slurry contains too much solid sample. Finally, it must be stressed that the introduction of samples with high organic contents in F-AAS can, in the long term, cause low sensitivity and poor precision.

2.7.2
US-SLE and US-SS for ET-AAS

ET-AAS is an instrumental technique that can be used for element determination for samples containing high levels of organic matter. In fact, element determination can be carried out directly in solids using ET-AAS. Two different approaches can be taken to join US-SLE and ET-AAS for element determination. The simplest is possible when total US-SLE can be attained. In such cases, after sonication, samples are allowed to stand while solid particles settle. The supernatant is then used to measure the content of the elements. Some organic matter is co-extracted during the sonication process, which can interfere with the measurement. The interferences caused are, however, minimal for most elements and are overcome with the use of adequate matrix modifiers [1, 18, 19, 40]. When elements are partially extracted from solids, the technique US-SS-ET-AAS can be used to introduce a mixture of the extracting solution plus solid particles into the graphite furnace. With this simple method total US-SLE is not mandatory. US-SS-ET-AAS gives rise, however, to more interferences than the introduction of the supernatant since considerably more organic matter is introduced into the graphite furnace [43]. Therefore, more time-consuming electrothermal programs are needed. In addition, the matrix modification employed needs to be modified in terms of the amount and type of matrix modifier used [43].

With the above approaches, Ag, Al, As, Ba, Be, Bi, Ca, Cd, Co, Cr, Cu, Fe, Ga, Hg, Mg, Mo, Mn, Ni, Li, Pb, Sb, Se, Si, Sn, Sr, Te, Ti, Tl, V and Zn have been determined in environmental, biological and inorganic samples [43] using ET-AAS.

2.7.3
US-SLE and US-SS for CV and HG Employed with AAS or AFS

CV generation for Hg and hydride generation, HG, for Pb, As, Se, Cd, Cu, Sn and Zn are versatile techniques widely used in analytical chemistry due to inherent simplicity of handling and low detection limits [44]. CV is carried out by first reducing the Hg^{2+} in solution using $SnCl_2$ or $NaBH_4$ to Hg^0, which is then stripped off from the solution using a flow of N_2 and transported to the measurement system, generally AAS or atomic fluorescence spectrometry (AFS). With HG, the ion in solution is transformed into its hydride, for example As(III) into AsH_3, which is then, like the Hg^0, stripped off with a flow of N_2 and transported to the measurement system. Ion transformation into hydrides is generally carried out using $NaBH_4$.

When US-SLE is used to extract mercury or any of the hydride-forming elements from environmental or biological samples, organic matter is co-extracted. This organic matter also reacts with the $SnCl_2$ or $NaBH_4$ used to transform the ions in solution. As a consequence, foaming is produced that makes measurements impossible, either by AAS or by AFS [10]. The simplest way to overcome this problem is to perform a pre-degradation step in the extract using the ultrasonic energy plus some oxidizing agents such as $KMnO_4$ or $HClO$ [9, 45]. In this way, the organic matter of the extract does not react with the $SnCl_2$ or the $NaBH_4$, thereby making measurement of the element possible. Another strategy consists of the centrifugation and filtration of the extract.

Importantly, regarding CV and HG methods, the element to be determined needs to be in solution in a known oxidation state, otherwise the concentration obtained will be lower than the actual one. As an example, mercury can be extracted from environmental samples either as inorganic mercury, Hg^{2+}, or organic mercury, for example, methylmercury, CH_3Hg^+. Taking into account that ultrasonic energy can transform methylmercury into inorganic mercury, depending on the reagents in solution, and that CV determination can only measure Hg^{2+}, when using $SnCl_2$ as reducing agent, a step to transform the organic mercury into an inorganic species is mandatory, unless a speciation study is being done [9]. Another example is the case of As. In oxidizing solutions, such as dilute HNO_3, the oxidation power of the ultrasonic energy tends to transform As(III) into As(V). This originates a problem since the kinetics of AsH_3 formation are much slower for As(V) than for As(III) [44]. Thus, the signal response for the same As concentration is lower for the higher oxidation state. For this reason, prior to total As determination by HG, a reduction step transforming As(V) into As(III) is mandatory after US-SLE. This step can be accomplished by using L-cysteine [10].

US-SS is an alternative method for element determination through Flow Injection-FI-CV(HG)-AAS that presents the advantages of reducing the time required for analysis, mercury losses and sample contamination. In this method, the slurry is formed using ultrasound for some seconds, then the ultrasonication is stopped and a representative aliquot composed of liquid and solid particles is injected into the FI system, as shown in Figure 2.2, with the injection valve (I) in the "load" position. The injection valve is changed to the "inject" position and then the sample is injected into the carrier stream. The reducing agent (R) and carrier (C) merge in the gas–liquid separator (GLS) where the Hg^0 is separated and transported using N_2 to a quart cell (Q), where it is detected by AAS [46]. This methodology, however, is markedly dependent on the action of the reducing agent on the analyte that is inside the solid particles, which severely limits the applicability of the approach. In addition, many

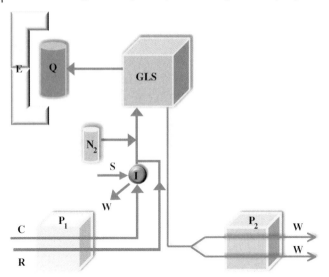

Figure 2.2 A manifold for US-SS-CV-AAS. C – carrier; R – reducing agent; S – sample as slurry; W – waste; I – injection valve; E – atomic absorption spectrometer; Q – quartz tube; GLS – gas–liquid separator; P1, P2 – peristaltic pumps.

variables should be optimized. For instance, too high a particle size may cause clogging of the injection valve/pumping system. Furthermore, aqueous calibration is not suitable for measurement as it is necessary to utilize the standard addition method [46]. US-SS-CV(HG)-AAS(AFS) has been used successfully in the determination of Hg, As, Sb, Se and Cd [47, 48].

2.8
Elemental Fractionation and Elemental Speciation

2.8.1
What is Speciation?

Elemental speciation and fractionation are analytical approaches that complement each other in terms of information.

The International Union for Pure and Applied Chemistry, IUPAC, define the fractionation of analytes as the "process of classification of an analyte or a group of analytes from a sample according to physical (e.g., size, solubility) or chemical (e.g., bonding, reactivity) properties."

By elemental speciation is meant any analytical methodology addressing information about the actual chemical form of an element rather than just its total amount in a sample. The definition of chemical species of an element is given by IUPAC as the "specific form of an element defined as to isotopic composition, electronic or oxidation state, and/or complex or molecular structure." Elemental speciation is important because of the differences in mobility and toxicity of the different element species. For example, As(III) and As(V) are both highly toxic, while the As organic forms present different degrees of toxicity, with arsenobetaine, the primary arsenic species found in many saltwater fish and shellfish, being non-toxic. As shown below, the identification and quantitative determination of the chemical forms of a metal in food or biota presents several challenges. The extraction and separation steps must be carefully considered to maintain the integrity of the species. Therefore, ultrasonication should be applied under chemical and physical conditions mild enough to liberate the species from the matrix, while avoiding its degradation or transformation. This requires elevated skill in the use of ultrasonic energy. Special care must be taken regarding interconversion of species, which can be caused by the oxidizing capacity of ultrasound. As explained in Chapter 1, the cavitation phenomena caused by ultrasonication promotes chemical reactions in the liquid media, such as the formation of oxidizing radicals (e.g., OH). The oxidizing environment caused by ultrasonication is enhanced when the extraction procedure is carried out in a solution containing oxidizing reagents such as HNO_3. The following examples show how this works. As(III) was almost entirely transformed into As(V) under the effects of an ultrasonic field (UP, 100 W, 5 min sonication time, 90% sonication amplitude) when the solvent was changed from 15% w/v HCl to 15% w/v HNO_3, meaning that extractions should be carried out in HCl if As species preservation is intended [10]. In another example, under the same conditions of ultrasonication it was observed that organomercurials were degraded to inorganic mercury in HCl solutions only if traces of HClO were also added [9]. More interestingly, the transformation of As(III) into As(V) has been described after 60 s of ultrasonication with a probe in a 1 M H_3PO_4 solution, while under the same conditions but with a shorter sonication time of 30 s no As interconversion was obtained [49]. Therefore, even if the protocol to be applied is obtained from literature, it is strongly recommended that the stability of the relevant species be studied before carrying out elemental speciation.

2.8.2
Shortening Sequential Fractionation Schemes

The different fractionation schemes described in the literature are based on the rational use of a series of more or less selective reagents, which are chosen to solubilize the different mineralogical fractions responsible for retaining the larger part of the elements. The idea behind fractionation schemes is to simulate the natural or man-caused conditions in the environment and see how those conditions affect the element release from soils and sediments to the environment [13]. As an example, acid rain caused by human activities can release a higher fraction of elements from soils than a less acidic rain. The fractions defined in a given scheme can vary according to the studied sample and author's thinking of what "fractionation" is [13]. Therefore, the European Community Standards, Measurement and Testing Programme, formerly BCR, launched in 1987 an attempt to harmonize fractionation schemes for trace-metal content in soils and sediments [50]. The BCR protocol entails three steps:

1. Extraction in acetic acid (0.11 M); fraction exchangeable bound to carbonates; called the "water and acid-soluble fraction;" extraction time 16 h.
2. Extraction in 0.5 M hydroxylamine acidified with 2 M HNO_3; fraction exchangeable bound to Fe-Mn oxides; called the "Reducible fraction;" extraction time 16 h.
3. Extraction in hydrogen peroxide (8.8 M); fraction exchangeable bound to organic matter and sulfides; called the "Oxidizable fraction;" extraction time 16 h.

As it can be seen, the drawback of fractionation schemes is the long time needed to complete the protocols. To speed up fractionation schemes different authors have highlighted the possibility of using ultrasonic energy to accelerate the different fractionation steps [51–53]. The first ultrasound-assisted sequential extraction procedure was carried out by applying ultrasound with a probe, using the reagents from the original BCR protocol in a sewage sludge sample, to successfully extract, in the respective sequential steps as defined above, Cu, Cr, Ni, Pb and Zn. The extractions in each fraction were compared with the standard BCR procedure for all the studied elements and no significant differences were found, except for Cr and Zn in the third fraction (oxidizable) [51]. After this pioneering work, ultrasound-assisted versions of the Tessier protocol, developed for both sewage sludge and river sediment, were soon reported for the same metals, showing some discrepancies for some of the fractions of the Tessier protocol [54, 55]. Different researchers have since then developed and applied ultrasound-assisted extraction to variants of both Tessier [56] and BCR [57, 58] extraction protocols. However, uncertainties remain about whether ultrasonication extracts elements in the same way as the standard procedure (agitation). Thus, along with the data claiming different extraction levels of Cr and Zn in the third (oxidizable) fraction, difficulties have also been stressed in obtaining similar fractionation patterns for matrix elements such as Fe by conventional and ultrasonic extraction procedures [52, 54]. It seems that the capacity of ultrasonication to accelerate sequential extraction schemes is directly linked to the matrix type and to the target element. Thus, the Tessier protocol has been successfully accelerated from 17 to 1.7 h with the aid of an UB for the element Sr in sediment samples [59]; and Cr, Mn, Fe, Co, Ni, Cu, Zn, As, Cd, Pb have been successfully extracted with an UP using a modified BCR procedure from mixed waste streams. The total time was reduced from the original 16 h to 15 min for each step [60].

Regarding sonication conditions, it seems that sonication times as short as possible and sonication frequencies at medium levels are recommended, trying always to avoid heating the solution as raising the temperature can change the equilibrium of the extracting solutions. Data reported at present are not conclusive about which sonication device should be used to speed up sequential extraction schemes, whether the UP or the UB. In addition, no interlaboratory trials have been developed to date to assess the performance of the accelerated protocols over the same matrix by different laboratories. Therefore, in the near future, more work dealing with this area is expected, focusing on the type and conditions by with sequential extraction schemes can be accelerated through ultrasonication. Table 2.3

Table 2.3 BCR sequential extraction scheme accelerated by different ultrasonic systems.

BCR sequential extraction steps	Treatment	Element recovery (%)	Reference
Step 1. Water and acid soluble fraction	UB, 43 kHz, 15 min, 25 °C	Cd (98 ± 20); Cu (104 ± 11); Fe (45 ± 8); Mn (97 ± 9); Zn (100 ± 11)	[112]
	UB, 35 kHz, 30 min	Cd (98.9); Cr (117.0); Cu (91.0); Pb (98.0); Ni (97.0); Zn (98.0)	[112]
	UB, 200 W, 3 h	Cu (ca 90); Zn(ca 70)	[52]
	UB, 20 min, 100%	Cd (ca 90); Cr (c. 80); Cu (c. 90); Ni (c. 90); Zn (c. 90)	[114]
	UP, 100 W, 20 kHz, 7 min	Cu (97.30); Cr(–); Ni (96.99); Pb(–); Zn (99.61)	[51]
	UP, 150 W, 3 min, continuous output	Cu (c. 100); Zn (c. 70)	[52]
Step 2. Reducible fraction	UB, 43 kHz, 15 min, 25 °C	Cd (–); Cu (97 ± 13); Fe (55 ± 6); Mn (78 ± 7); Zn (102 ± 13)	[112]
	UB, 35 kHz, 30 min	Cd (96.2); Cr (95.0); Cu (90.0); Pb (97.0); Ni (97.0); Zn (98.0)	[113]
	UB, 200 W, 1 h	Cu (c. 70); Zn (c. 60)	[52]
	UB, 20 min, 100%	Cd (c. 100); Cr (c. 90); Cu (c. 90); Ni (c. 90); Zn (c. 95)	[114]
	UP, 100 W, 20 kHz, 7 min	Cu (96.62); Cr (–); Ni (ND); Pb (–); Zn (96.58)	[51]
	UP, 150 W, 5 min, pulsed output	Cu (ca 80); Zn (ca 60)	[52]
Step 3. Oxidizable fraction	UB, 43 kHz, 15 min, 25 °C	Cd (–); Cu (68 ± 9); Fe(86 ± 11); Mn (–); Zn (100 ± 14)	[112]
	UB, 35 kHz, 30 min	Cd (99.0); Cr (95.0); Cu (91.0); Pb (98.1); Ni (97.0); Zn (97.0)	[113]
	UB, 200 W, 1 h	Cu (c. 70); Zn (c. 90)	[49]
	UB, 2 min + 6 min, 70%	Cd (c. 100); Cr (c. 100); Cu (c. 94); Ni (c. 100); Zn (c. 100)	[114]
	UP, 100 W, 20 kHz, 2 min + 6 min	Cu (96.94); Cr (96.01); Ni (96.84); Pb (97.80); Zn (97.87)	[51]
	UP, 150 W, 1 min, continuous output	Cu (c. 80); Zn (c. 40)	[52]

shows the type of ultrasonication device used, the conditions of sonication time and sonication amplitude, along with the type of extraction scheme and the results obtained.

2.8.3
Speciation for Soils and Sediments

Literature reports of the speciation of elements in soils and sediments through ultrasonic extractions are still scarce, and are related to a few elements. Arsenic speciation has been carried out in soils and sediments, the species studied being As (III), As(V), dimethylarsinic acid (DMA) and monomethylarsonic acid (MMA). The extractions were performed with UP in H_3PO_4 1 M [49] and phosphate buffer at pH 5.5–6.5 [61] in short times of 30 and 60 s, respectively. Both procedures reported centrifugation and filtration prior to arsenic species separation, through High-Performance Liquid Chromatography, HPLC, and As determination by ICP-MS and HG-AFS, respectively. Sb(III) has been distinguished from Sb(V) in road soil after extraction in an UB during 45 min. The extracting agent was citric acid at pH 2.58. Centrifugation followed by filtration was the cleaning procedure prior to HPLC-isotope dilution inductively coupled plasma mass spectrometry (ID-ICP-MS) determination [62]. Six different organotin compounds have been also extracted from sediment with an UP using a sonication time of 30 s and a sonication amplitude of 60% in a (1:9) solution of acetic acid : methanol. A cleaning step consisting of centrifugation at 3000 rpm during 5 min was carried out prior to Sn determination through gas chromatography with an atomic emission detector, GC-AED [63]. Butyltin compounds have been reported as being highly extractable from sediments, by using either UP (15 min sonication time, extracting solvent: methanol : acetic acid 1 : 3 v/v) [64] or UB (4 min sonication time, extracting agent glacial acetic acid) [65]. As can be seen, there is a lack of work regarding speciation of metals in soils and sediments using ultrasonication as the methodology for enhancing the SLE. For this reason an increasing number of publications on this subject is expected in the near future.

2.8.4
Speciation from Plants

Allium sativum and *Brassica juncea* plants have been grown in a medium rich in Se and the extraction efficiency for Se and the Se species integrity studied in three different extracting solutions, (i) a solution containing the enzyme protease, (ii) a 0.1 M HCl solution and (iii) a 25 mM ammonium acetate buffer, pH 5. 6. The extractions were performed with an UP using a sonication time of 3 min. Total Se extraction efficiencies were in the order protease solution > buffer solution > HCl solution, with the recoveries in all experiments ranging from 75 to 122%; no significant differences were found when applying enzymatic hydrolysis of the plant tissues instead of acid or buffer extractions. The most important finding of the study was that the use of UP allows a faster extraction of the Se species (using HCl or acetate buffer as extracting solutions), with no species degradation. This could be an extremely important approach for future speciation studies. With protease extraction, the samples were first passed through a membrane of pore size 3000 Da to eliminate the excess protease that might disturb the chromatographic separation. Even so, the chromatographic profile obtained by reverse phase ion pair (RP-IP)-HPLC for the protease extraction was completely different to those observed for HCl and buffer solutions. Accordingly, no speciation studies regarding protease solution and ultrasonication were performed [66].

Speciation studies have been carried out on rice straw shoots and rice straw roots (Oryza sativa) using ultrasonication for 30 min in an UB in the following solutions: water, methanol, water–methanol (1 : 1, v/v), water–ethanol (1 : 1, v/v) and water–acetonitrile (1 : 1, v/v). The most effective solvent for extraction of the arsenic species was water–acetonitrile (1 : 1). However, the extraction efficiency was only 49%, which is lower than the extraction obtained with microwave energy (85%). This data could suggest that for some speciation studies the more powerful UP instead of the UB should be used [67]. The combination of enzymes and ultrasonication with probe, recently reported [11, 12], has become the most promising combination to perform chemical speciation in plants. As a general trend, when using enzymes and ultrasonication, no organics solvents have to be used, sonication times are short and low amounts of sample and reagents are needed, thereby matching the concept of analytical minimalism [68]. Speciation in rice using enzymes in conjunction with ultrasonication with a probe has been carried out as follows: quantitative extraction for total arsenic (>95%) and 90% extraction for the sum of the arsenic species [As(III), As(V), methylarsonic acid (MMA) and dimethyarsinic acid (DMA)] without species transformation in only 3 min by applying an aqueous mixture of protease XIV and α-amylase [69].

Speciation of Se has been achieved in selenium-enriched lentil plants (Lens esculenta), by extraction using a UP with a 3-mm titanium microtip at 20 W power and 20 kHz sonication frequency, according to the following procedure: 25 mg of sample + 10 mg of protease XIV + 3 mL of water + sonication during 2 min. The mixture was centrifuged (7500g, 30 min) using a 10 kDa cut-off filter. Total Se and Se(IV), Se(VI), SeMetO, SeMetSeCys and SeMet were measured [70]. Selenium and Se-species have been also determined in yeast in a work that suggests that the enzymatic treatment accelerated by US can extract the total content of Se in times as short as 5 s, but longer ones, 30 s, are needed if Se speciation is required. This could suggest that total Se is extracted to solution by the joint action of enzyme and ultrasound in a short time, but that more time is needed to degrade the macromolecules that contains the Se-amino acids, if speciation is intended [39]. In this case the procedure was: (i) For total metal determination: 10 mg of sample + 1 mg of protease XIV + 1 mL water + sonication for 5 s at 20 W; the mixture was centrifuged and 20 µL of the supernatant was diluted up to 10 mL with water. (ii) For metal speciation: 10 mg of yeast + 1 mg of protease XIV + 1 mL of water sonication for 30 s. The mixture was filtered through a 0.45 µm nylon filter.

An interesting finding regarding the improvement in element speciation obtained by the application of enzymes and ultrasound in plants is given by the differences in extractability reported for the above-mentioned As speciation in rice using a solution of water–acetonitrile (1 : 1) and UB and the work reporting As speciation in rice using enzymes and UP [69]. Although the As species obtained for both sample treatments were the same, As(III), As(V), MMA and DMA, the sample treatment using enzymes provided an As recovery of 99.7% against the low 49% reported by using the extracting solution water–acetonitrile (1 : 1).

2.8.5
Speciation from Soft Tissues

Two approaches are currently used for longer sonication times from soft tissues. The first uses ultrasonic energy and solutions containing different chemicals, whilst the second uses ultrasonic energy and enzymes. As noted throughout this chapter, ultrasonic energy can change the oxidation state of elements or promote element species counterchange. This is why most workers prefer not to use ultrasonic energy to speed up element speciation. Nevertheless, it is becoming clear that new findings on element speciation, especially those entailing enzymes, involve the use of ultrasonic energy.

Concerning the speciation of an element with ultrasonication and non-enzymatic approaches, studies reported in the literature are linked to the three elements Hg, As and Se.

For Hg speciation, alkaline extractions have been used to determine mercury and methylmercury in fish samples. For this purpose, extraction efficiencies with UB (3 h sonication time, 75 °C, 50 W L^{-1}), UP (25 min sonication time, 100 W potency at a 100% sonication amplitude) and microwave energy (6 min at 60 W potency) have been compared [71]. The extracting solution was 18% (w/v) NaOH in methanol. All the extraction systems provided good results; the measurements were carried out by solid-phase microextraction (SPME)-GC-AFS, and the extractions were done in tuna fish. The aforementioned results confirm that Hg extractions using UB should be performed with the aid of heating and the use of longer sonication times than with the UP. The report does not mention MeHg$^+$ artifact formation, one of the main problems when dealing with mercury speciation [72]. Further confirmation of the utility of alkaline extractions is given by the MeHg$^+$ speciation carried out in sea-food in conjunction with the speciation of MBT (monobutyltin), DBT (dibutyltin) and TBT (tributyltin) by GC-AES [73]. In this work, extraction and derivatization were carried out in separate steps. Sonication was performed for 1 h in an UB in a 2 M KOH solution, with the recoveries being about 100%. Other approaches take advantage of the ultrasonic energy to enhance not only the extraction process but also the derivatization, performing both extraction and derivatization at the same time under the effects of an ultrasonic field. For example, following this approach, MeHg$^+$ has been determined in biological samples [74], such as lobster hepatopancreas, dogfish liver, tuna fish and pike. The procedure was based on acid sample leaching (5 min shaking) followed by simultaneous *in situ* derivatization and further extraction (40 min sonication time, UB) in the presence of sodium tetraethylborate and nonane, buffered at pH 7.0. An overall average recovery of methylmercury of >90% was obtained, with analysis by GC coupled to microwave induced plasma atomic emission spectrometry, MIP-AES, or ICP-MS.

Acid extractions accelerated by ultrasonication have been also reported as a valuable approach in speciation studies. Notably, however, the literature provides a wide variety of options to perform mercury speciation in soft tissues through acid extractions. One conclusion that can be drawn from literature deals with the special care that must be taken with the concentrations/combinations of the acids used to carry out the extraction. Once again it must be stressed that conditions must be such as to allow element extraction while maintaining species integrity. This is illustrated in the following example: Hg and MeHg$^+$ have been determined after acid extraction in sea-food [75], in a work in which different combinations and concentrations of the reagents HCl, HNO$_3$, H$_2$SO$_4$ and H$_2$O$_2$ were assayed. The study concluded that reliable results are obtained only with the following combination/concentrations: 3 M HCl + 1.4 M HNO$_3$ + 0.11 M H$_2$SO$_4$ and 0.22 M H$_2$O$_2$. Extractions were made with an UB (50 W potency, 50 kHz sonication frequency). In the aforementioned method, a fully mechanized procedure was developed for the speciation of mercury in fish samples using CV-AFS. Sample slurries in the optimum acid mixture, in the presence of a surfactant and with traces of K$_2$Cr$_2$O$_7$, were injected into a flow system, sonicated and merged with 1 mL of an oxidizing mixture of KBr/KBrO$_3$ heated at 50 °C in a water bath. The same sonicated sample slurries were also measured, in the absence of KBr/KBrO$_3$, to obtain a second series of data that could be employed to establish the concentrations of free Hg(II). By difference the methyl-mercury was obtained. One advantage of acid extractions aided by ultrasonication is that they allow one to distinguish between inorganic mercury and methylmercury due to selective extraction of the target element species as a function of acid concentration. Thus, a simple method using HCL for mercury speciation in fish tissues has been proposed [76]. Centrifuged extracts were injected directly into a FI-CV-AAS system. First, methylmercury was determined separately after selective extraction with 2 M HCl. Second, inorganic mercury was determined by selective reduction with stannous chloride in 5 M HCl extracts containing both mercury species. Total mercury could not be determined in the sonicated acid extracts using sodium tetrahydroborate as reducing agent because the methylmercury and inorganic mercury sensitivities were different. Total mercury was in agreement with that determined in the samples after MWPAD.

Reports on As speciation without the use of enzymes have stressed the convenience of using ultrasonication with probe instead of UB, to obtain higher extraction efficiencies in shorter times. As an example [77], As(III), As(V), MMA and DMA have been extracted from algae samples and determined by HPLC-ICP-AES using three extracting methods, namely, magnetic stirring, UB and UP. For each system, four extracting solutions were tested: tris(hydroxymethyl)aminomethane, THAM, H$_3$PO$_4$, H$_2$O and different mixtures of H$_2$O/CH$_3$OH. Two interesting conclusions can be drawn from this study. First, the level of extractable arsenic was the same for the three methods of extraction, with the soluble As accounting for 65% of the total arsenic in the sample. Interestingly, the best As recoveries were obtained in H$_2$O. Second, both magnetic stirring and the UB needed an extraction time of 30 min while extraction with UP required only 30 s. The same authors subsequently developed the use of microwave energy as a tool for extracting As at 90 °C during 5 min in water from the same matrixes [78]. It was claimed that the latter conditions afford a better As extraction than that obtained with UP. The authors also stressed As species preservation during all of the extraction treatments studied. Other work remarks, however, that neither ultrasonic energy nor microwave energy alone can efficiently extract As and As species; instead, both systems need to be combined for some soft tissues [79]. For example, As speciation was achieved in chicken tissue, using three energetic systems to speed up the extraction process, UB, UP and microwave energy, and three extracting solutions were assayed, H_2O, H_2O/CH_3OH (1:1 v/v) and Cl_3CH/CH_3OH (1:1 v/v). Interestingly, the individual use of each extracting protocol was not sufficient for quantitative recovery of As(III), DMA, MMA, As(V) and AsB in spiked samples of chicken tissue. Therefore, an extraction procedure using a microwave oven and then an UP, carrying out the extractions in a H_2O/CH_3OH (1:1 v/v) solution was done, thereby enabling the extraction of the arsenic species in 7 min with efficiencies ranging from 80 to 100%. HPLC-ultraviolet UV-HG-AFS was used with success to measure the As species. The procedure was 100% efficient when applied to real samples of chicken tissue, in which AsB and nitarsone were detected. Some non-chromatographic As speciation studies have also been reported. These studies are dedicated more to differentiating the sum of As(V) + As(III) or toxic arsenic from the less toxic species AsB. Here, toxic As is the sum of As(III), DMA, MMA and As(V). For the first case [80], the differentiation of As(V) and As(III), a SS-FI-HG-GF-AAS method for the determination of As(III) and total inorganic As, without total sample digestion, was developed for As speciation in fish samples. The method involves trapping arsenic vapor on a pre-heated graphite furnace inner wall at 300 °C, treated with 150 mg of iridium as permanent chemical modifier. The sample was treated in two steps. First an extraction process was carried out using an UP in an HCl media (60% sonication amplitude, 40 W potency, 1 min sonication time), and the resultant slurry was purged with ozone (0.14 mg min^{-1}) to oxidize the organic matter and the species present in solution. To estimate the As(III) and As(V) concentrations in samples, the difference between the analytical sensitivities of the absorbance signals obtained for arsenic hydride with and without previous treatment of samples with thiourea was used. Thiourea transforms As(V) into As(III). The concentration of As(V) was calculated as the difference between total As and As(III). This method has the advantage that only a minimum of reagents and sample handling are required, thereby reducing the risks of contamination and/or analyte loss; in addition, chromatographic speciation approaches are not necessary. However, calibration is achieved by the technique of standard additions. It must be also considered that approaches calculating some element species indirectly, by subtracting signals or concentrations among species, lack precision as a function of the amount of each species. Namely, if one species X_1 is at low level and its concentration is indirectly quantified by subtracting the concentration of a second species X_2 to another third one X_3, then the standard deviation, SD, of the calculated concentration for the species X_1, SDX_1, will be the square root of the sum of the square of the standards deviations of the measurements of X_2, SDX_2 and X_3, SDX_3 [i.e., $(SDX_1)^2 = (SDX_3)^2$

$(SDX_2)^2$]; consequently, sometimes the SD of the calculated concentration will be higher than its actual value. To differentiate toxic As from the less toxic species AsB, understanding toxic As to be the sum of As(III), As(V), DMA, and MMA, as mentioned above, extractions were carried out in lobster hepatopancreas, dogfish muscle and dogfish liver, using an UP (90% sonication amplitude and 5 min sonication time) in a HCl 15% m/v solution. After centrifugation at 1800 rpm for 10 min, the supernatant was sonicated under a flow of ozone, transforming DMA and MMA into inorganic As. Then, all As(V) in solution was transformed into As(III) by using L-cysteine as reducing agent, and it was measured by FI-HG-AAS. The method was appropriate for determining total As in solutions containing the species As(V), As(III), DMA and MMA. AsB can not be determined using this methodology.

For Se speciation, a remarkable work in which five different extracting solutions were tested for total Se determination and speciation has been reported [81]. The extracting solutions studied were (i) mineralization with $HNO_3 + H_2O_2$ to give total content; (ii) enzymatic hydrolysis to yield $92 \pm 1\%$ Se extraction (10 mg sample/ 100 mg protease, medium phosphate/citric acid buffer with a 7.5 pH, stirring for 24 h in water bath at 37 °C); (iii) warm water extraction (100 mg sample) with 5 h shaking to yield 20% of total Se; (iv) organic extraction (100 mg sample, 2 : 3 : 5 water : chloroform : methanol) with 5 h shaking to yield $11 \pm 2\%$ Se extraction; and (v) acid hydrolysis (50 mg + 1.5 mL HCl 6 M, stirred for 5 h in a water bath) to yield $8 \pm 0.2\%$ Se extraction. Major selenium species determined were selenomethionine (42%) and selenocysteine (35%). As can be seen, good extractions were only obtained using enzymes. Long extraction times, however, remained the main limitation for speciation purposes, as shown in the above-mentioned work: 24 h for enzymatic hydrolysis, 5 h for warm water extraction, 5 h for organic extraction and 5 h acid extraction. Acceleration of the extracting procedures through the use of ultrasonic energy seems to fail for the extraction of Se, unless enzymes are also used. As an example, Table 2.4 shows different extraction schemes accelerated with ultrasonic energy, by either UB or UP. As can be seen, total extraction, while preserving species integrity, in times as short as 30 s is only obtained using enzymes plus UP. Remarkably, buffering the solution to maintain the best enzyme performance was not necessary to obtain good results [39].

Since it was reported to the scientific community in 2004, ultrasonic assisted enzymatic digestion (USAED) using an UP has become a routine method for element speciation [11, 12, 39]. Nevertheless, the methodology is still being developed. Although enzymes were previously described in the literature as a tool for chemical speciation, as explained above, never before had the enzymatic reaction kinetics been speeded up using an UP for element speciation purposes. Previously, the enzymatic hydrolysis was accelerated by means of UBs or heating baths at 37 °C. However, this "acceleration" required from overnight to hours to be completed. Enzymatic hydrolysis in seconds or a few minutes can be done using only the ultrasonic energy provided by an UP. Studies on the enhancement of enzymatic kinetics by ultrasonic energy have linked the enhancement with the intensity of sonication rather than with the frequency of sonication [82]. The question of how ultrasonic energy enhances enzymatic activities remains contro-

Table 2.4 Total Se and Se species recovery from different samples using USLE in the sample treatment.

Aim	Sample	Treatment	Sample mass/volume (mg/mL^{-1})	Se recovery (%)	Reference
Total Se	Seafood	UB, 35 min, 90 °C, 2.5 M HNO$_3$	200/6	100	[40]
Total Se	Seafood	Probe sonication, 3 min, 50% amplitude, 0.5 M HNO$_3$	20/1.5	100	[7]
Se speciation	White clover CRM 402	UB, 2 × 30 min, methanol/water (1 + 1)	500/10	28.5 ± 3.5	[115]
		UB, 2 × 30 min, methanol/water (1 + 1) + 0.28 M HCl		36.7 ± 0.3	
Se speciation	Yeast	UB, 2 × 30 min, methanol/water (1 + 1) + 4% M HCl		47.6 ± 5.5	
		UB, 60 min, methanol/water (1 + 9) + 0.2 M HCl	200/5	13	[116]
Se sequential Extraction	Yeast	UB, 60 min, 30 mM Tris-HCl buffer (pH 7) + 0.1 mM PMSF		10	
		Probe sonication, 10 × 5 s, 10 mM Tris-HCl buffer (pH 8)		15.2 ± 1.0	[117]
Total Se yeast		Probe sonication, 30 s, 0.8 M HCl		19.5 ± 0.4	[39]
		Probe sonication, 240 s, 0.8 M HCl	20/10	21 ± 3	
		Probe sonication, 480 s, 0.8 M HCl	20/10	19 ± 1	
		Probe sonication, 30 s, 0.01 M HCl	20/10	14 ± 1	
		Probe sonication, 240 s, 0.01 M HCl	20/10	16 ± 4	
		Probe sonication 30 s, 250 mL methanol + 750 mL of water	20/10	21 ± 4	
		Probe sonication 240 s, 250 mL methanol + 750 mL of water	20/10	23 ± 6	
		Probe sonication 480 s, 250 mL methanol + 750 mL of water	20/10	14 ± 1	
		Probe sonication 480 s, 250 mL methanol + 750 mL of 0.8 M HCl in water	20/10	23 ± 4	
		Probe sonication 480 s, 250 mL methanol + 750 mL of 1.2 M HCl in water	20/10	19.4 ± 0.1	
		Probe sonication, 30 s, methanol	20/10	17 ± 10	
		Probe sonication, 240 s, methanol	20/10	14 ± 1	
		Probe sonication, 120 s, water	10/1	32 ± 8	
		Probe sonication, 30 s, enzyme subtilisin + 0.1 M Tris-HCl buffer (pH 7.5)	20/5	101 ± 7	
		Probe sonication, 90 s, 0.1 M Tris-HCl buffer (pH 7.5)	10/1	91 ± 5	
		Probe sonication, 30 s, enzyme protease XIV, 0.1 M Tris-HCl buffer (pH 7.5)	10/1	104 ± 8	
				98 ± 16	

versial. It has been suggested that there is an increased contact area between phases caused by cavitation, allowing a reduction of mass transfer limitations in the enzyme–substrate system. In addition, the reduction in sample size caused by high intensity focused ultrasound (i.e., particle disruption) allows more substrate area to be in contact with the enzyme per second of sample treatment. Furthermore, the enhancement in reaction rates caused by ultrasonication has been attributed to an increase in collisions between enzyme and substrate [83]. As consequence of the aforementioned phenomena, it seems that ultrasonic cavitation promotes an increment in enzymatic reaction rates rather than a change in enzymatic reaction constants. As an example, under the effects of an ultrasonic field, the maximum reaction rate (V_{max}) of the hydrolysis promoted by the enzyme lipase was increased while the Michaelis constant (K_m) remained unaltered [84]. On the other hand, it has been reported that while short sonication times enhance enzymatic hydrolysis long sonication times lead to enzyme inactivation [84].

For the correct application of USAED some variables of the analytical procedure must be carefully considered [12]. The correlation between samples and enzymes is one of the most important. The enzymes used for element speciation belong to the family of hydrolases, consisting of lipases, amylases and proteases. Lipases hydrolyze fats, amylases hydrolyze starch and glycogen, and proteases hydrolyze proteins and peptides. Therefore, not all hydrolases can be used for the same matrix, and even for the same family of hydrolases differences in the extraction efficiencies have been reported. Thus, three proteases have been studied in conjunction with ultrasonication with probe for the extraction of Se from mussel tissue. Under the same conditions of sample treatment, the Se extractions were (i) $93 \pm 7\%$ for protease XIV, (ii) $70 \pm 9\%$ for subtilisin A and (iii) $28 \pm 5\%$ for trypsin [23]. The use of cocktails of enzymes to improve the extraction of elements from plant or animal tissues has also been described. For instance, (α-amylase + protease type XIV) and (protease XIV + lipase) have been used for As speciation in rice and hair respectively [50, 85, 86]. For some type of samples a multistep enzymatic digestion is the only way to obtain accurate results. For example, for a botanical sample, the enzyme amylase is first used to break the wall of the cells, made mainly of starch and glycogen. Through this first step the protein content of the sample is liberated into the liquid media, and then the speciation is completed in a second step, in which a different type of enzyme, a protease, is used to digest the proteins, freeing the element species associated with them. pH and temperature are also important variables to take into account for the correct application of USAED. Regarding pH, published data suggest that the activities of enzymes having their optimum activities at pH values around 7 can be boosted with ultrasonication even if the extracting solution is not buffered at such a pH [12, 39]. This finding is important since it facilitates chromatographic separations and helps to avoid sample contamination caused by buffer solutions. However, in some cases, buffering the solution can be critical to obtain accurate results. Thus, compared with a non-buffered media, a Se extraction 28% higher was obtained when protease XIV was used in a buffered media for Se speciation in chicken samples [87]. In terms of temperature, some experiments have revealed that performing USAED under cooling conditions leads to lower recoveries than obtained at room temperature, thus showing that a compromise between the optimum temperature for an efficient enzyme activity and the optimum for a good ultrasonic cavitation to occur must be obtained [12]. As a general rule, a continuous sonication time at room temperature with an UP for a short time, 2–3 min, is recommended for USAED treatment. Another, critical, issue is the substrate/enzyme ratio. The lower the ratio, the higher the amount of enzyme used. This makes the analysis expensive. In addition, the higher the total amount of sample for a given volume of solution in an ultrasonic extraction, the higher the amount of organic matter co-extracted to solution. A high amount of organic matter in solution is a problem for numerous analytical techniques, since it can hinder the analyte signal. However, by doubling the amount of enzyme used in Se extractions from chicken muscle, liver and kidney, the recoveries were increased from 86 to 97%, 84 to 93% and 81 to 95%, respectively [87]. For this reason, a study of metal extraction as a function of the amount of enzyme used for a fixed quantity of sample is recommended when USAED is applied for the first time in a new matrix. Concerning enzyme ageing, the extraction efficiency of protease XIV for the extraction of Se from mussel tissue fell by 20% after 3 months [12]. For this reason, it is recommended that enzymes be purchased in as small amounts as possible to avoid storage and that, once the enzyme container is opened, the enzyme must be used as soon as possible. The type of sonication device used to perform USAED is one of the most important variables to be considered. Since the UB, unlike the UP or the SR device, can not boost enzymatic hydrolysis in short times, 2–4 min, it is not recommended for use in USAED. Finally, as a general rule, USAED performed with the aid of an UP requires sonication times of 2–5 min and low sonication amplitudes, generally 20–30% [12]. A comprehensive scheme for the rapid application of USAED in the laboratory is presented in Figure 2.1, where it can be seen that sample mass used should be between 10 and 50 mg – this value depends mainly on the limit of detection of the analytical technique, on the target analyte concentration in the sample and on sample homogeneity. Higher amounts of sample and volumes can also be used. The sample size should be as low as possible. In this way, the total area in contact with the enzyme is increased. A comparison between the amount of metal extracted for buffered and non-buffered aqueous solutions should always be done, since pH can affect enzyme activity. If buffered solutions are required to obtain accurate results, the choice of buffer used must take into account the subsequent speciation process through HPLC. Total element extraction with USAED methodology must always be compared with a classic method for total element extraction, such as MWPAD. When the use of enzymes for element extraction gives low recoveries, some changes can be made to the sample treatment to increase the levels of metals extracted. For instance, the sonication time or the amount of enzyme can be increased. A graph of percentage of metal extracted versus amount of enzyme added for a fixed amount of sample must always be plotted. Other approaches include choosing another type of enzyme or trying an enzyme cocktail or a sequential enzymatic extraction using different enzymes. If the amount of metal extracted is still low, the analytical criteria of the chemist must be used to decide if the concentration is high enough to justify further research regarding chemical speciation. Once it has been proved that the extraction is accurate for more than

80% of the total expected metal extracted, with a reasonable standard deviation (<10%), the next step is to ensure that there is no species interconversion due to the sample treatment. To do this, the samples must be spiked with standards of the element species, such as DMA or MMA for As, and subjected to the sample treatment. Recoveries will indicate whether there are transformations. Once it has been demonstrated that (i) more than 80% of the element is extracted and that (ii) there are no species interconversion, the USAED methodology can be applied for element speciation.

The chemical forms of As that have been found after USAED sample treatment in rice [49, 69] and human hair [85] are As(III), As(V), arsenochlorine (AsC), arsenobetaine (AsB), dimethyl-arsonic acid (DMA) and monomethylarsonic acid (MMA) and As(III), As(V), DMA and MMA respectively. It must be emphasized that no As-species interconversion was claimed to occur by the authors.

Most data available in literature concerning USAED are dedicated to Se, which has been studied in yeast [39], chicken [87] and Antarctic krill [88]. Selenomethionine (SeMet) has been found in yeast, SeMet in chicken and Antarctic krill, and Se (IV), Se(VI), selenomethionine Se-oxide (SeMetO), SeMetSeCys and SeMet have been found in lentil [70]. The stability of Se species under the effects of an ultrasonic field provided by an UP, in the presence of enzymes, has been stressed, so that neither species degradation nor species interconversion was described for this metal. However, Pedrero et al., using USAED, have pointed out that Se speciation, done by anion exchange chromatography, corresponding to stem and roots from plants grown in the presence of selenite and selenate, was not able to discriminate between selenocysteine (SeCys2) and SeMetO as both co-eluted at a retention time of 2.1 min. Consequently, a second chromatography column, combining ion exchange and size exclusion mechanisms, had to be used for unambiguous species identification [70].

2.8.6
Speciation from Other Types of Samples

Ultrasonication can be used to perform element speciation, not only in solid samples but also in liquid ones. These methods are based on non-chromatographic approaches, and are done mainly by AAS or AFS. For this reason, this type of speciation can not provide the same information that a systematic chromatographic approach does. Thus, speciation in liquid samples using ultrasonic energy has been limited to elements with mainly two different species in solution, such as Hg and $MeHg^+$, or to elements for which species degradation is easily obtained through ultrasonication. For instance, DMA and MMA are easily transformed into inorganic arsenic [either As(III) or As(V)] by applying ultrasonic energy in a solution containing an oxidant agent such as $KMnO_4$. Let us see how this works. Toxic arsenic in human urine [As(III) + As(V) + MMA + DMA] has been determined by FI-HG-AAS in two steps [89]. First, MMA and DMA were degraded using $KMnO_4$ and HCl to inorganic As [As(III) + As(V)] and then all As in solution was transformed into As (III) using a solution of the reducing agent L-cysteine. This protocol has two advantages. The first is that at the same time that MMA and DMA are degraded the organic matter present in the urine is also degraded. This degradation is mandatory, otherwise the formation of AsH_3 is not possible due to chemical and physical interferences caused by the organic matter. The second advantage comes from the selective As species degradation of the protocol: AsB is not degraded. Since AsB does not form a hydride but As(III) does, both species can be separated by HG-AAS. In conclusion, As can be differentiated by the described protocol in urine as toxic As [As(III) + As(V) + MMA + DMA] and AsB, an almost non toxic species, which is the most abundant one in fish and shellfish.

Organic mercury species present in water and urine can be distinguished from inorganic mercury by CV-AAS following a chemical scheme similar to the one described above for As speciation. Complete oxidation of $MeHg^+$ to Hg^{2+} was accomplished within 90 s whilst phenyl-Hg and diphenyl-Hg were degraded within 10 s using a 50% sonication amplitude (100 W nominal UP power, 20.5 kHz sonication frequency) and 1 M HCl liquid medium in the presence of traces of ClO^-. Taking advantage of the selective Hg^{2+} reduction using $SnCl_2$, which only converts Hg^{2+} into Hg^0, the speciation scheme is simple: first the Hg^{2+} content (inorganic fraction) is measured in the sample without ultrasonic treatment, then the sample is submitted to ultrasonic treatment and all organic species are transformed into Hg^{2+}, and thereby the total amount of mercury can be measured. The mercury organic fraction is obtained by subtracting from the total Hg the inorganic Hg fraction. This methodology or similar ones have been reported for Hg speciation in urine, drinking waters, seawaters and wastewaters [8, 9, 90].

The carcinogenic Cr(VI) has been determined in workplace air filter samples immediately after collection of the samples. The sample treatment has been described as being carried out with the aid of an UB [91]. This method uses ultrasonic extraction (US-SLE) to extract Cr(VI) from the sampling media, and strong anion-exchange solid-phase extraction (SAE-SPE) to separate Cr(VI) from Cr(III) and other interferences. Apparently, there is no counter-conversion between Cr(VI) and Cr(III) caused by UB. Air filter samples were removed from the sampling cassettes and placed in 15-mL centrifuge tubes using PTFE-coated forceps. Next, 10 mL of 0.10 M Na_2CO_3/0.02 M $NaHCO_3$ (pH 10.7 $\pm$ 0.1) extraction buffer was added to each centrifuge tube. The centrifuge tubes containing the filter samples were then immersed in the UB and subjected to sonication for 30 min at ambient temperature. The water level in the bath was adjusted so that it was above the level of sample solution inside the centrifuge tubes. Portable spectrophotometric measurement of Cr(VI) was then conducted using the 1,5-diphenylcarbazide method [92].

Sb(III) and Sb(V) have been determined in environmental samples after an extraction procedure aided by ultrasonic energy. The extraction was carried out in a citric acid solution as follows: 0.1 g of airborne particulate matter was weighed into a screw-top polypropylene bottle (20 mL), and then 10 mL of citric acid (26 mM) was added. The extraction was performed in an ultrasonic water bath, for 30 min at room temperature. The resultant Sb(III)- and Sb(V)-citrate complexes were separated on a PRP-X100 anion-exchange column with 10 mM EDTA/1 mM phthalic acid (pH 4.5) as a mobile phase. The extract was then filtered through a 0.45 mm membrane prior to measurements by HPLC-ICP-MS [93].

2.9
On-Line Applications

On-line applications for element analysis using ultrasonication are becoming a useful tool in analytical chemistry, since these approaches avoid sample contamination, allowing, at the same time, high sample throughput and low reagent consumption. Different combinations of on-line approaches are represented in Figures [2.3–2.5].

It must be stressed, however, that on-line applications of ultrasonic energy are severely limited when carried out in an indirect form, that is, when ultrasonic energy is not applied directly to the sample. As explained in Chapter 1, the ultrasonic efficiency in terms of cavitation is drastically diminished when the wave must cross the wall of the sample container. This problem is frequently found in the application of ultrasonic energy with UBs. Therefore, on-line applications can not be generalized and must be always assayed for each target analyte and for each matrix of interest.

In on-line applications of ultrasound two methodologies can be found. In the first, a solid is introduced in a column, which is then placed inside an UB or in a heated water bath. In the latter case ultrasonication is provided by an UP situated over the column (Figure 2.3). The solid introduced into the column can be the sample or a dedicated solid for specific purposes, such as, for example, a resin for liquid–solid element separation/preconcentration. In the second methodology, the on-line approach entails the application of ultrasonic energy to a liquid solution to, for example, enhance chemical reactions such as organic matter degradation. In both cases, the system can be adapted to work in a continuous mode or with a stop-flow. The latter permits a longer sample treatment over the entire bulk of sample.

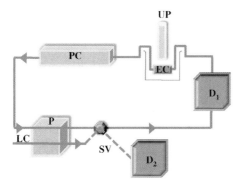

Figure 2.3 On-line approaches for ultrasonic applications. The sample is introduced in the column (EC), which is immersed in a UB or a water bath. In the latter case ultrasonication is provided by an ultrasonic probe (UP) situated over the column. The solution is pumped by the peristaltic pump (P). If the solution is an extracting one a preconcentration (OC) column can be also incorporated. The extracting process can be monitored using a detector (D1), for instance a UV detector. Once the process has been completed a second detector (D2) can be used to monitor the element, for instance a AA spectrometer.

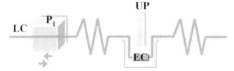

Figure 2.4 Open on-line approaches for ultrasonic applications. The sample is introduced in the column (EC), which is immersed in a UB or a water bath. In the latter case ultrasonication in provided by an ultrasonic probe (UP) situated over the column. The solution is pumped by the peristaltic pump (P). The pump can be stopped so that the solution remains under the effects of the ultrasonic field for longer. The solution can also be pumped in directly in an inverse direction through the column.

2.9.1
Open and Closed Systems

When the ultrasonic energy is applied in an on-line procedure without recirculation, the system is defined as "open system" (Figure 2.4) [94]. In this approach the flow of

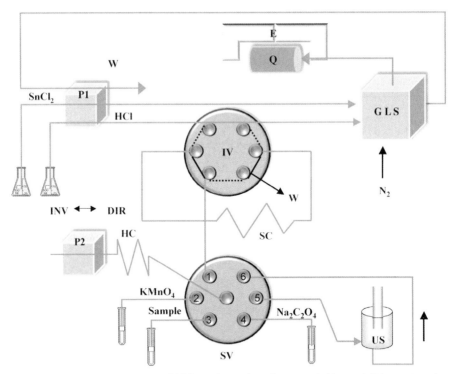

Figure 2.5 SIA/HIFU/FIA manifold for on-line determination of mercury by cold vapor atomic absorption spectroscopy. Key: (sample) sample solution mixed with HCl and NaClO; (P1 and P2) peristaltic pumps; (INV) inverse flow; (DIR) direct flow; (HC) holding coil; (IV) injection valve with a 500 µL sample loop; (SV) selection valve; (W) waste; (Q) quartz cell, (E) atomic absorption spectrometry; (GLS) gas–liquid separator; (N_2) stream of N_2; (US) UP.

the liquid can be stopped to maintain the effects of the ultrasonic field on the sample of interest for longer times. If the on-line application is done with recirculation, the system is called a "closed system" (Figure 2.3) The advantage of the recirculation system is that it allows monitoring of the process for which ultrasonic energy is being used. When the process is finished the flow can be re-directed towards the detector.

2.9.2
UB

Owing its simplicity of use in on-line systems requiring ultrasonic energy, the UB is by far the most widely used ultrasonic device for this purpose. As some examples, the successful extraction has been reported with on-line applications of (i) Fe and Cu from mussel tissue [95, 96], (ii) Mn from solid and seafoods [97] and (iii) Fe from meat [98]. One of the most interesting applications of on-line approaches with UBs involves integrating the extraction process with a subsequent element preconcentration procedure in the same on-line scheme. This has allowed, for example, the determination of Cd and Pd in mussel samples by F-AAS [99]. The use of dilute nitric acid as extracting agent in conjunction with an UB, at a flow rate of 3.5 mL min^{-1} at room temperature, was sufficient for quantitative extraction of these trace metals. A minicolumn containing a chelating resin (Chelite P, with aminomethylphosphoric acid groups) proved to be an excellent material for the quantitative preconcentration of Cd and Pb prior to their detection. A good precision for the whole procedure (2.0 and 2.3%), high enrichment factors (20.5 and 11.8) and detection limits of 0.011 and 0.25 µg g^{-1} for Cd and Pb, respectively, were obtained for 80 mg mussel samples.

However, as well as its proven applicability for element analysis, some difficulties in terms of efficiency have been also described in the literature for on-line approaches using liquid–solid element preconcentration in resins with subsequent solid–liquid extraction aided by UBs. As an example, two different sonication procedures, one on-line and the other off-line, have been assessed to implement a reliable solid–liquid extraction of Hg from commercial resins (Dowex and chelex-100) previously used to concentrate Hg^{2+} from urine [90]. The urine had undergone prior treatment with an UP in conjunction with $KMnO_4$ and HCl, allowing organic matter degradation in less than 3 min. $95 \pm 10\%$ of mercury in certified urine and $97 \pm 6\%$ of the spiked methylmercury were recovered with the Dowex resin plus the static ultrasonic procedure, while $96 \pm 11\%$ of the spiked mercury was recovered with the Dowex resin plus the dynamic procedure for which, however, ultrasonication was not necessary. Hence, on-line approaches must be checked carefully for adequate performance, especially if resins are involved in the analytical treatment. As explained above, when the on-line system does not work properly, the bath procedure can also be tried with success.

2.9.3
UP

To the best of our knowledge the UP has been used mainly for indirect ultrasonic applications (Figure 2.3) for on-line systems. This was the method used to extract Ca,

Mg, Fe, Cu and Zn from animal feeds [100], Cd and Pb from plants [38] and Cr(VI) from soils [101]. However, as in the case of UB, indirect application of ultrasonic energy can limit the applicability of on-line approaches due to the decrease of cavitation efficiency as a consequence of the wall of the sample container. Thus, although the extraction of Hg retained in a column previously filled with Dowex resin was attempted with different solutions of HNO_3 and ultrasonication, the latter provided with an UB and an UP at the same time, quantitative Hg recovery from the resin was only possible using high concentrations of HNO_3, with the ultrasonication being unnecessary [93].

A recent interesting advance uses UP on-line with direct application on the sample [102]. This new concept is based on the coupling of on-line ultrasonication with probe with a sequential injection/flow injection analysis (SIA/FIA) system. Although the method has been used for mercury speciation in waters and urine, it is expected to be applied in future to other kinds of samples such as solids transported as slurries.

2.10
Current Trends

2.10.1
Accelerating Liquid–Liquid Extractions

On-line approaches making use of the advantages of ultrasonication for metal extraction from a liquid phase to an organic one or vice versa have been also reported [103, 104]. Certainly, simple applications for element liquid–liquid extractions can be applied without requiring complicated schemes. As an example, an interesting application uses extraction in an organic solvent and back extraction to water solution aided by ultrasonication for Hg separation, preconcentration and back-extraction from human urine as follows [105]: the urine is first oxidized with $KMnO_4$/HCl/focused ultrasound (6 mm probe diameter). Secondly, the mercury is extracted and preconcentrated with dithizone and cyclohexane. Finally, the mercury is back-extracted from the organic solution to water and preconcentrated again with the aid of focused ultrasound (3 mm probe diameter). The procedure allowed mercury determination by ET-AAS with fast furnace analysis and calibration against aqueous standards. Matrix modification was provided by the $KMnO_4$ used in the sample treatment. The procedure was accomplished with a low sample volume (8.5 mL) and in less than 3 min per sample. The preconcentration factor attained was 14. Recoveries from spiked urine with inorganic mercury, methyl-, phenyl- and diphenyl-mercury ranged from 86 to 98%.

When trying to implement ultrasonication to speed up or to improve element liquid–liquid extraction the following must be taken into account: (i) ultrasonic energy does not always improve liquid–liquid extraction between two immiscible phases; (ii) ultrasonic energy applies better to extractions from organic phases to aqueous ones than vice versa; (iii) care must be taken to promote extractions while avoiding chemical reactions caused by ultrasonication . In the last case, however, a chemical reaction can also be pursued concurrently with the extraction procedure. As an example, in the above-mentioned case of mercury pre-concentration from human urine, the back-extraction procedure includes the destruction of the complex formed in the organic phase between the mercury and the dithizone.

2.10.2
Chemical Vapor Formation

A new method for the measurement of Hg that does not need reducing agents such as $SnCl_2$ or $NaBH_4$ based on cold vapor formation and ultrasonic energy has been reported [106]. The method entails the sonication of Hg in a liquid solution containing an organic acid such as formic acid. The mechanism seems to lie in the reduction of Hg^{2+} to Hg^0 by reducing gases formed by the sonolysis of the solvent followed by Hg^0 volatilization due to the degassing effect of ultrasonic energy. The proposed mechanism responsible for the cold vapor generation promoted by ultrasonic irradiation is as follows:

$$2HCOOH \rightarrow CO_2 + H_2 + CO + H_2O$$

$$Hg^{2+} \rightarrow Hg^0_{sol}$$

$$Hg^0_{sol} \rightarrow Hg^0_{gas}$$

This new application opens up new fields for total mercury determination and speciation in liquid samples. In addition, the possibilities of on-line approaches using sonochemical vapor formation is anticipated.

2.11
Conclusion

Ultrasonic energy is a powerful tool for elemental determination and chemical speciation, including the acceleration of sequential extraction schemes. Although ultrasonic energy can not be universally applied to extract all types of elements in all type of matrixes, it can be used to extract many of them from a huge number of sample types. In addition, the combination of ultrasound and enzymes opens up new analytical strategies for chemical speciation and for protein and peptide identification through mass spectrometry-based techniques. Finally, the accomplishment of ultrasound and on-line systems anticipates a promising future of automation for the different applications dealt with in this chapter.

Abbreviations

AFS	Atomic fluorescence spectroscopy
DBT	Dibutyltin

DMA	Dimethylarsinic acid
ET-AAS	Electrothermal atomic absorption spectroscopy
F-AAS	Flame atomic absorption spectroscopy
FI	Flow injection
FIA	Flow injection analysis
FI-CV-AAS	Flow injection, cold vapor atomic absorption spectroscopy
GC-AED	Gas chromatography with atomic emission detector/gas chromatography with microwave-induced plasma atomic emission spectroscopy
HG	Hydride Generation
HPLC	High performance liquid chromatography
HPLC-ID-ICP-MS	High performance liquid chromatography-isotope dilution inductively coupled plasma mass spectrometry
HPLC-UV-HG-AFS	High performance liquid chromatography ultra-violet hydride generation atomic fluorescence spectroscopy
ICP-MS	Inductively coupled plasma mass spectrometry
MBT	Monobutyltin
MMA	Monomethylarsonic acid
MWPAD	Microwave pressure acid digestion
SeMet	Selenomethionine
SeMetO	Selenomethionine Se-oxide
SeMetSeCys	Selenomethionine-selenocysteine
SES	Sequential extraction scheme
SIA	Sequential injection analysis
SM&T	Standard, measurement and testing program
SS-FI-HG-GF-AAS	Slurry sampling flow injection hydride generation graphite furnace atomic absorption spectroscopy
TBT	Tributyltin
UB	Ultrasonic bath
US-SLE	Ultrasonic extraction
UP	Ultrasonic probe
US	Ultrasounds
USAED	Ultrasonic assisted enzymatic digestion
US-SLE	Ultrasonic solid–liquid extraction
US-SS-ET-AAS	Ultrasonic slurry sampling electrothermal atomic absorption spectroscopy

References

1 Capelo, J.L., Maduro, C. and Vilhena, C. (2005) *Ultrasonics Sonochemistry*, **12**, 225–232.

2 Maduro, C., Vale, G., Alves, S. *et al.* (2006) *Talanta*, **68**, 1156–1161.

3 Vaisanen, A. and Suontamo, R. (2002) *Journal of Analytical Atomic Spectrometry*, **17**, 739–742.

4 Capelo, J.L. and Ana Mota (2005) *Current Analytical Chemistry*, **2**, 193–201.

References

5. Capelo, J.L., Ximenez-Embun, P., Madrid-Albarran, Y. and Camara, C. (2004) *Trends in Analytical Chemistry*, **23**, 331–340.
6. Ashley, K., Andrews, R.N., Cavazos, L. and Demange, M. (2001) *Journal of Analytical Atomic Spectrometry*, **16**, 1147–1153.
7. Mendez, H., Alava, F., Lavilla, I. and Bendicho, C. (2002) *Analytica Chimica Acta*, **452**, 217–222.
8. Capelo, J.L., Lavilla, I. and Bendicho, C. (2000) *Analytical Chemistry*, **72**, 4979–4984.
9. Capelo, J.L., Rivas, G.M., Oliveira, L.G. et al. (2006) *Talanta*, **68**, 813–818.
10. Capelo, J.L., Lavilla, I. and Bendicho, C. (2001) *Analytical Chemistry*, **73**, 3272–3736.
11. Bermejo, P., Capelo, J.L., Mota, A. et al. (2004) *Trends in Analytical Chemistry*, **23**, 654–663.
12. Vale, G., Rial-Otero, R., Mota, A. et al. (2008) *Talanta*, **75**, 872–884.
13. Gleyzes, C., Tellier, S. and Astruc, M. (2002) *Trends in Analytical Chemistry*, **21**, 451–467.
14. Tessier, A., Campbell, P.G.C. and Bisson, M. (1979) *Analytical Chemistry*, **51**, 844–851.
15. Quevauviller, P.H., Rauret, G. and Griepink, B. (1993) *International Journal of Environmental Analytical Chemistry*, **51**, 231–235.
16. Filgueiras, A.V., Lavilla, I. and Bendicho, C. (2002) *Analytical and Bioanalytical Chemistry*, **374**, 102–108.
17. Capelo, J.L., Galesio, M.M., Felisberto, G.M. et al. (2005) *Talanta*, **66**, 1272–1280.
18. Amoedo, L., Capelo, J.L., Lavilla, I. and Bendicho, C. (1999) *Journal of Analytical Atomic Spectrometry*, **14**, 1221–1226.
19. Capelo, J.L., Lavilla, I. and Bendicho, C. (1998) *Journal of Analytical Atomic Spectrometry*, **13**, 1285–1290.
20. Lima, E.C., Barbosa, F., Krug, F.J. et al. (2000) *Journal of Analytical Atomic Spectrometry*, **15**, 995–1000.
21. Santos, H.M. and Capelo, J.L. (2008) *Talanta*, **73**, 198–205.
22. Pena-Farfal, C., Moreda-Pineiro, A., Bermejo-Barrera, A. et al. (2004) *Analytical Chemistry*, **76**, 3541–3547.
23. Vale, G., Pereira, S., Mota, A. et al. (2007) *Talanta*, **74**, 198–205.
24. Milher-Ihli, N.J. (1995) *Spectrochimica Acta*, **50**, 447–488.
25. Mierzwa, J., Sun, Y.C. and Yang, M.H. (1998) *Spectrochimica Acta*, **53**, 63–69.
26. Felipe-Sotelo, M., Carlosena, A., Andrade, J.M. et al. (2005) *Microchemical Journal*, **82**, 217–224.
27. dos Santos, E.J., Hermann, A.B., Frescura, V.L.A. and Curtius, A.J. (2005) *Analytica Chimica Acta*, **548**, 166–173.
28. Hristozov, D., Domini, C.E., Kmetov, V., Stefanova, V., Georgieva, D., Canals, A. (2004) *Analytica Chimica Acta*, **516**, 187–196.
29. Segade, S.R., Albor, M.C.D., Gomez, E.F. and Lopez, E.F. (2003) *International Journal of Environmental Analytical Chemistry*, **83**, 343–356.
30. Nascentes, C.C., Korn, M. and Arruda, M.A.Z. (2001) *Microchemical Journal*, **69**, 37–43.
31. Mierzwa, J., Sun, Y.C., Chung, Y.T. and Yang, M.H. (1998) *Talanta*, **47**, 1263–1270.
32. Dobrowolski, R. and Mierzwa, J. (1993) *Fresenius' Journal of Analytical Chemistry*, **346**, 1058–1061.
33. Milher-Ihli, N.J. (1993) *Fresenius' Journal of Analytical Chemistry*, **345**, 482–489.
34. Milher-Ihli, N.J. (1990) *Fresenius' Journal of Analytical Chemistry*, **337**, 271–274.
35. Caballo-López, A. and Luque de Castro, M.D. (2007) *Talanta*, **71**, 2074–2079.
36. Maurii-Amejo, A.R., Arnandis-Chover, T., Marín-Saez, R. and Llobat-Estellés, M. (2007) *Analytica Chimica Acta*, **581**, 78–82.
37. Filgueiras, A.V., Lavilla, I. and Bendicho, C. (2001) *Fresenius' Journal of Analytical Chemistry*, **369**, 451–456.
38. Ruiz-Jiménez, J., Luque-García, J.L. and Luque de Castro, M.D. (2003) *Analytica Chimica Acta*, **480**, 231–237.

39 Capelo, J.L., Ximenez-Embun, P., Madrid-Albarrán, Y. and Camara, C. (2004) *Analytical Chemistry*, **76**, 233–237.

40 Bermejo-Barrera, P., Muniz-Naveiro, O., Moreda-Pineiro, A. and Bermejo-Barrera, A. (2001) *Analytica Chimica Acta*, **439**, 211–227.

41 Lavilla, I., Mosquera, A., Millos, J. et al. (2006) *Analytica Chimica Acta*, **566**, 29–36.

42 Ferreira, H.S., Dos Santos, W.N.L., Fiúza, R.P. et al. (2007) *Microchemical Journal*, **87**, 128–131.

43 Cal-Prieto, M.J., Felipe-Sotelo, M., Carlosena, A. et al. (2002) *Talanta*, **56**, 1–51.

44 Lajunen, L.H.J. and Peramaki, P. (2004) *Spectrochemical Analysis by Atomic Absorption and Emission*, 2nd edn, The Royal Society of Chemistry, Cambridge.

45 Capelo, J.L., Maduro, C. and Mota, A.M. (2006) *Ultrasonics Sonochemistry*, **13**, 98–106.

46 Rio-Segade, S. and Bendicho, C. (1999) *Journal of Analytical Atomic Spectrometry*, **14**, 1907–1912.

47 dos Santos, E.J., Hermann, A.B., Frescura, V.L.A. and Curtius, A.J. (2005) *Analytica Chimica Acta*, **548**, 166–173.

48 Matusiewich, H. (2003) *Applied Spectroscopy Reviews*, **38**, 263–294.

49 Sanz, E., Munoz-Olivas, R. and Camara, C. (2005) *Journal of Chromatography. A*, **1097**, 1–8.

50 Quevaiviller, P.H., Rauret, G., Muntau, H. et al. (1994) *Fresenius' Journal of Analytical Chemistry*, **349**, 808–814.

51 Perez-Cid, B., Lavilla, I. and Bendicho, C. (1998) *Analytica Chimica Acta*, **360**, 35–41.

52 Davidson, C.M. and Delevoye, G. (2001) *Journal of Environmental Monitoring*, **3**, 398–403.

53 Bacon, J.R. and Davidson, C.M. (2008) *Analyst*, **133**, 25–46.

54 Perez-Cid, B., Lavilla, I. and Bendicho, C. (1999) *Fresenius' Journal of Analytical Chemistry*, **363**, 667–672.

55 Perez-Cid, B., Lavilla, I. and Bendicho, C. (1999) *International Journal of Environmental Analytical Chemistry*, **73**, 79–92.

56 Vaisanen, A. and Kiljunen, A. (2005) *International Journal of Environmental Analytical Chemistry*, **85**, 1037–1049.

57 Canepari, S., Cardarelli, E., Ghigni, S. and Scimonelli, L. (2005) *Talanta*, **66**, 1122–1130.

58 Krasnodebska-Ostrega, B., Kaczorowska, M. and Golimowski, J. (2006) *Mikrochimica Acta*, **154**, 39–43.

59 Elik, A. and Akay, M. (2001) *International Journal of Environmental Analytical Chemistry*, **80**, 257–267.

60 Greenway, G.M. and Song, Q.J. (2002) *Journal of Environmental Monitoring*, **4**, 950–955.

61 Huerga, A., Lavilla, I. and Bendicho, C. (2005) *Analytica Chimica Acta*, **534**, 121–128.

62 Amereih, S., Meisel, T., Scholger, R. and Wegscheider, W. (2005) *Journal of Environmental Monitoring*, **7**, 1200–1206.

63 Campillo, N., Aguinaga, P., Vinas, I., Lopez-Garcia, M. and Hernandez-Cordoba (2004) *Anal. Chim. Acta*, **525**, 272–280.

64 Ruiz-Encinar, J., Rodríguez-Gonzalez, P., Garcia-Alonso, J.I. and Sanz-Medel, A. (2002) *Analytical Chemistry*, **74**, 270–281.

65 Carpinteiro, J., Rodríguez, I. and Cela, R. (2001) *Fresenius' Journal of Analytical Chemistry*, **370**, 872–877.

66 Montes-Bayon, M., Diaz-Molet, M.-J., Gonzalez, E.B. and Sanz-Medel, A. (2006) *Talanta*, **68**, 1287–1293.

67 Yuan, C.G., Jiang, G.-B. and He, B. (2005) *Journal of Analytical Atomic Spectrometry*, **20**, 103–110.

68 Halls, D.J. (1995) *Journal of Analytical Atomic Spectrometry*, **10**, 169–175.

69 Sanz, E., Munoz-Olivas, R. and Camara, C. (2005) *Analytica Chimica Acta*, **535**, 227–235.

70 Pedrero, Z., Encinar, J.R., Madrid, Y. and Camara, C. (2007) *Journal of Chromatography. A*, **1139**, 247–253.

71 Abrankó, L., Kmellar, B. and Fodor, P. (2007) *Microchemical Journal*, **85**, 122–126.

72 Leermakers, M., Baeyens, W., Quevauviller, P. and Horvat, M. (2005) *Trends in Analytical Chemistry*, **24**, 383–393.

73 Zabaljauregui, M., Delgado, A., Usobiaga, A. *et al.* (2007) *Journal of Chromatography. A*, **1148**, 78–85.

74 Tu, Q., Quiana, J. and Frech, W. (2000) *Journal of Analytical Atomic Spectrometry*, **15**, 1583–1588.

75 Cava-Montesinos, P., Dominguez-Vidal, A., Cervera, M.L. *et al.* (2004) *Journal of Analytical Atomic Spectrometry*, **19**, 1386–1390.

76 Rio-Segade, S. and Bendicho, C. (1999) *Journal of Analytical Atomic Spectrometry*, **14**, 263–268.

77 Garcia-Salgado, S., Quijano-Nieto, M.A. and Bonilla-Simon, M.M. (2006) *Talanta*, **68**, 1522–1527.

78 Garcia-Salgado, S., Quijano Nieto, M. and Bonilla-Simon, M.M. (2006) *Journal of Chromatography. A*, **1129**, 54–60.

79 Sánchez-Rodas, D., Gómez-Ariza, J.L. and Oliveira, V. (2006) *Analytical and Bioanalytical Chemistry*, **385**, 1172–1177.

80 Matusiewicz, H. and Mroczkowska, M. (2003) *Journal of Analytical Atomic Spectrometry*, **18**, 751–761.

81 Gilon, N., Astruc, A., Astruc, M. and Potin-Gautier, M. (1995) *Applied Organometallic Chemistry*, **9**, 623–628.

82 Sakakibara, M., Wang, D., Takahasshi, R. and Mori, S. (1996) *Enzyme and Microbial Technology*, **18**, 444–448.

83 Bracey, E., Stenning, R.A. and Brooker, B.E. (1998) *Enzyme and Microbial Technology*, **22**, 147–151.

84 Talukder, M.M.R., Zaman, M.M., Hayashi, Y. *et al.* (2006) *Biocatalysis and Biotransformation*, **24**, 189–194.

85 Sanz, E., Muñoz-Olivas, R., Dietz, *et al.* (2007) *Journal of Analytical Atomic Spectrometry*, **22**, 131–139.

86 Mihuez, V.G., Tatar, E., Virag, I. *et al.* (2007) *Food Chemistry*, **105**, 1718–1725.

87 Cabañero, A.I., Madrid, Y. and Camara, C. (2005) *Analytical and Bioanalytical Chemistry*, **381**, 373–379.

88 Siwek, M., Noubar, A.B., Bergmann, J. *et al.* (2006) *Analytical and Bioanalytical Chemistry*, **384**, 244–249.

89 Correia, A., Galesio, M., Santos, H. *et al.* (2007) *Talanta*, **72**, 968–975.

90 Patricio, A., Fernandez, C., Mota, A.M. and Capelo, J.L. (2006) *Talanta*, **69**, 769–775.

91 Hazelwood, K.J., Drake, P.L., Ashley, K. and Marcy, D. (2004) *Journal of Occupational and Environmental Hygiene*, **1**, 613–619.

92 National Institute for Occupational Safety and Health (NIOSH) (2003) Method 7703, Hexavalent Chromium by Portable Visible Spectrophotometry, in *NIOSH Manual of Analytical Methods*, 4th edn (3rd Suppl.) (eds P.C., Schlecht and P.F., O'Connor), NIOSH, Cincinnati, Ohio.

93 Zheng, J., Iijima, A. and Furuta, N. (2001) *Journal of Analytical Atomic Spectrometry*, **16**, 812–818.

94 Priego-Capote, F. and Luque de Castro, M.D. (2004) *Trends in Analytical Chemistry*, **23**, 644–652.

95 Yebra, M.C. and Moreno-Cid, A. (2002) *Journal of Analytical Atomic Spectrometry*, **17**, 1425–1428.

96 Moreno-Cid, A. and Yebra, M.C. (2002) *Spectrochimica Acta Part B-Atomic Spectroscopy*, **57**, 967–974.

97 Yebra, M.C. and Moreno-CId, A. (2003) *Analytica Chimica Acta*, **477**, 149–155.

98 Moreno-Cid, A., Yebra, M.C., Cancela, S. and Cespón, R.M. (2003) *Analytical and Bioanalytical Chemistry*, **377**, 730–734.

99 Yebra-Biurrun, M.C., Cancela-Perez, S. and Moreno-Cid-Barinaga, A. (2005) *Analytica Chimica Acta*, **533**, 51–56.

100 Priego-Capote, F. and Luque de Castro, M.D. (2004) *Analytical and Bioanalytical Chemistry*, **378**, 1376–1381.

101 Luque-Garcia, J.L. and Luque de Castro, M.D. (2002) *Analyst (Cambridge, UK)*, **127**, 1115–1120.

102 Fernandez, C., Conceicão, Antonio, C.L., Rial-Otero, R. *et al.* (2006) *Analytical Chemistry*, **78**, 2494–2499.

103 Luque de Castro, M.D. and Priego-Capote, F. (2007) *Talanta*, **72**, 321–334.

104 Ruiz-Jimenez, J. and Luque de Castro, M.D. (2003) *Analytica Chimica Acta*, **489**, 1–9.
105 Capelo, J.L., dos Reis, C.D., Maduro, C. and Mota, A. (2004) *Talanta*, **64**, 217–223.
106 Gil, S., Lavilla, I. and Bendicho, C. (2006) *Analytical Chemistry*, **78**, 6260–6264.
107 Epstein, M.S., Carnrick, G.R., Slavin, W. and Miller-Ihli, N.J. (1989) *Analytical Chemistry*, **61**, 1414–1419.
108 Fernandez-Cortes, C., Lavilla, I. and Bendicho, C. (2006) *Spectroscopy Letters*, **39**, 713–725.
109 Vaisanen, A. and Ilander, A. (2006) *Analytica Chimica Acta*, **570**, 93–100.
110 Dos Santos, E.J., Herrmann, A.B., Frescura, V.L.A. and Curtius, A.J. (2005) *Journal of Analytical Atomic Spectrometry*, **20**, 538–543.
111 Furukawa, M. and Tokunaga, S. (2004) *Journal of Environmental Science and Health Part A-Toxic/Hazardous Substances & Environmental Engineering*, **39**, 627–638.
112 Dabek-Zlotorzynska, E., Kelly, M., Chen, H. and Chakrabarti, C.L. (2003) *Analytica Chimica Acta*, **498**, 175–187.
113 Kazi, T.G., Jamali, M.K., Siddiqui, A. *et al.* (2006) *Chemosphere*, **63**, 411–420.
114 Ipolyi, I., Brunori, C., Cremisini, C. *et al.* (2002) *Journal of Environmental Monitoring*, **4**, 541–554.
115 Emtebog, H., Bordin, G. and Rodríguez, R.A. (1998) *Analyst*, **123**, 245–253.
116 Casiot, C., Szpunar, J., Lobinski, R. and Potin-Gautier, M. (1999) *Journal of Analytical Atomic Spectrometry*, **14**, 645–650.
117 Encinar, J., Ruzik, R., Buchmann, W. *et al.* (2003) *Analyst*, **128**, 220–224.

3
Ultrasonic Assisted Extraction for the Analysis of Organic Compounds by Chromatographic Techniques
Raquel Rial-Otero

3.1
Introduction

Different extraction techniques can be used to extract volatile and semivolatile organic compounds from liquid and solid matrixes. Traditional techniques are normally time consuming, tedious, require the use of large amounts of toxic organic solvents and glassware and can be relatively expensive. Therefore, recent trends in sample preparation have focused on the development of methods with the following characteristics:

- simplicity, preferably one step;
- low cost, by reduction of time and solvent consumption;
- miniaturization by reduction of sample size;
- environmentally friendly, by reduction of the use of toxic organic solvents;
- possibility of automation [1].

Recently, the ultrasonic assisted extraction (UAE) or ultrasonic solvent extraction (USE) has been introduced as a new sample preparation technique for the extraction of organic compounds from liquid and solid matrixes. This technique is based on the use of ultrasonic energy to ensure a more efficient contact between the sample and the extraction solvent, allowing rapid extractions of organic compounds from liquid and solid matrixes. On this basis, and due to its simplicity, UAE has gained in popularity in recent years, becoming a powerful technique for the extraction of organic and inorganic compounds. In fact, the number of publications in the last eight years that include the words "ultrasonic assisted extraction" is 233, about 74% of which were published between 2004 and 2007 (Figure 3.1). In addition, some reference methods for organics extraction are also based on the power of ultrasound. For example, the United States Environmental Protection Agency (USEPA), in its method 3550, proposes the use of UAE for the extraction of organic compounds at low concentration levels from solids such as soils, sludges and wastes [2].

Ultrasound in Chemistry: Analytical Applications. Edited by José-Luis Capelo-Martínez
Copyright © 2009 WILEY-VCH Verlag GmbH & Co. KGaA, Weinheim
ISBN: 978-3-527-31934-3

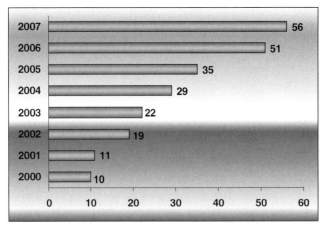

Figure 3.1 Evolution of the number of publications based on ultrasonic assisted extraction during recent years. Source ISI Web of Knowledge (v.4.1) database.

3.2
Overview of Classic and Modern Extraction Procedures for Organics

Conventionally, organic compounds can be extracted by liquid–liquid extraction (LLE) or solid phase extraction (SPE) in the case of liquid samples and by solid–liquid partitioning (SLP) or Soxhlet in the case of solid samples. However, as noted above, these methods are time consuming, tedious, require too much organic solvents and can be relatively expensive. Moreover, their application may result in analyte losses, contamination and poorer precision. In addition, Soxhlet extraction is generally not suitable for volatile organic compounds (VOCs) from solid samples, due to large VOC losses when refluxing the solvent during the typical long extraction times of this technique. In addition, further losses occur if concentration of the extract is required. An exhaustive explanation of the classic extraction methods used for organic compounds is beyond scope of this chapter. For a complete understanding of these techniques, the book *Sample Preparation Techniques in Analytical Chemistry* is highly recommended [3]. In this book the reader will find the basic concepts of these techniques, practical information about their implementation and also a review of numerous published applications.

Modern extraction techniques for organics have been developed recently with the aim of reducing some of the drawbacks connected with classic extraction techniques. Some of the new modern methodologies are: supercritical fluid extraction (SFE), pressurized liquid extraction (PLE), also called accelerated solvent extraction (ASE), microwave assisted extraction (MAE), static headspace analysis (HS), dynamic headspace analysis or purge-and-trap (PT), solid phase microextraction (SPME) and stir bar sorptive extraction (SBSE). Figure 3.2 shows schematized diagrams of these processes. Many of them have proven to be superior to traditional extraction

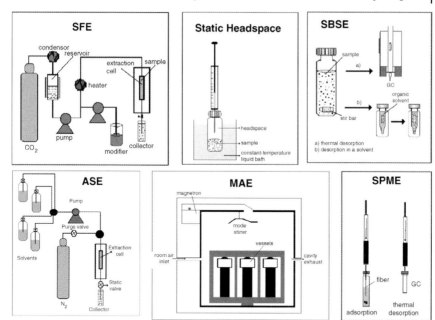

Figure 3.2 Schematized diagrams of some modern extraction techniques for organics: supercritical fluid extraction (SFE); accelerated solvent extraction (ASE); microwave assisted extraction (MAE); static headspace analysis (HS); solid phase microextraction (SPME); and stir bar sorptive extraction (SBSE).

techniques, in terms of factors such as solvent consumption, sample size, extraction time and analyte recovery. These techniques are reviewed briefly below, focusing on their major advances and drawbacks.

Supercritical fluid extraction (SFE) uses the unique properties of supercritical fluids, which are substances above their critical temperature and pressure, to facilitate the extraction of organics from solid samples. The sample is loaded into an extraction cell and placed in an oven. The extract is collected either by a sorbent trap or by a collection vial containing a solvent. For optimum recoveries, temperature, pressure, flow rate and extraction time should be optimized. This method has been approved by the USEPA for the extraction of total petroleum hydrocarbons, polycyclic aromatic hydrocarbons (PAHs), organochlorine pesticides (OCPs) and polychlorinated biphenyls (PCBs) [2].

Accelerated solvent extraction (ASE) uses conventional solvents at elevated temperatures (100–180 °C) and pressures (1500–2000 psi) to enhance the extraction of organics from solids. The combination of elevated pressures and temperatures affects the solvent, the sample and their interactions. For example, high pressure allows the solvent to penetrate deeper into the sample matrix, and at higher temperatures analyte solubility increases, solvent viscosity and surface tension are reduced and the mass transfer is faster [3]. ASE has been applied to the extraction of organic compounds from different samples and it has also been approved as a standard method by the

USEPA for the extraction of water insoluble or slightly water soluble semivolatile organic compounds, such as organophosphorus pesticides (OPPs), OCPs, chlorinated herbicides, PCBs, polychlorinated dibenzodioxins and polychlorinated dibenzofurans in environmental samples such as soils, clays, sediments, sludges and waste solids [2].

Indeed, ASE and SFE yield good recoveries in a short time for some semivolatile and nonvolatile organic compounds when extracted from solid samples. However, there are some drawbacks [4]:

1. They are generally not suitable for VOCs determinations.
2. In both techniques, losses of volatile analytes might occur during the extracting collection step (hot solvent discharge to a vented flask in ASE or supercritical fluid decompression in SFE).
3. Their routine use in sample preparation is considered to be expensive due to the cost of the instrumentation required.
4. SFE often requires the addition of organic modifiers.
5. Successful analyte recovery depends on, among other factors, the type of matrix [3].

Consequently, a better understanding of the matrix effects is necessary before these techniques can be used as standard extraction methods [1].

Microwave assisted extraction (MAE) was introduced in 1986 and is based on the use of organic solvents to extract organic pollutants from solids, using microwave energy as a heating source. In conventional heating, thermal energy is transferred from the source to the object through conduction and convection. However, heating during MAE is based on the effects of microwaves on molecules by ionic conduction and dipole rotation [3]. MAE efficiency can be influenced by several factors, such as extraction solvent, temperature, extraction time, matrix effects and water content. This extraction technique has been applied successfully to extract organic compounds, such as PAHs, pesticides, PCBs and phenols, from various solid and liquid matrixes, such as sediments, soils, plant materials and water samples. In addition, it has been approved by the USEPA as a standard method for the extraction of semivolatile and nonvolatile compounds from soil samples. MAE is more effective and less expensive than many of the new methods, and it reduces the sample preparation time and solvent volumes substantially in comparison with conventional extraction techniques. Nevertheless, it is not very popular. This may be because its effectiveness is highly dependent on matrix characteristics [4]. In addition, for safety reasons, MAE requires the use of special microwave systems designed for organic analysis [5].

Static headspace extraction (HS) and dynamic headspace extraction or purge-and-trap (PT) are the most popular techniques for the analysis of volatile organic compounds. VOCs are organic compounds whose vapor pressures are greater than or equal to 0.1 mmHg at 20 °C. With static headspace extraction, the sample is placed in a sealed headspace vial that is introduced in an oven or in a liquid bath at constant temperature. Thus, volatile analytes diffuse into the headspace of the vial until equilibrium is reached and, then, a portion of the headspace sample is introduced into a gas chromatograph for analysis [3]. Equilibrium is usually difficult to achieve and, therefore, high temperatures are required to increase the diffusivity and the gas/solid distribution of the volatile analytes, contributing in this manner to analyte losses. In addition, quantitative analysis is very difficult in static headspace and it may be necessary to optimize the extraction process to obtain good sensitivity and accuracy [4]. When an exhaustive extraction is required or trace quantities of analytes must be analyzed, dynamic headspace extraction is preferred over static headspace extraction. In this case, a purge gas passes through the sample continuously, providing a concentration gradient that aids in the exhaustive extraction of analytes, which are trapped by a sorbent. Once the purge is completed, analytes are thermally desorbed from the trap into a gas chromatograph for analysis [3]. This method is recommended by USEPA for the analysis of volatile contaminants that have diffused onto external surfaces but not into internal micropores of solid matrixes. Consequently, this method is not suitable for analysis of VOCs in clay samples [4].

Solid phase microextraction (SPME) is a solvent-free extraction technique, originally developed by Pawliszyn and coworkers [6]. SPME uses a fiber rod (fused silica or metal) coated with a polymeric coating to extract organic compounds from their matrixes. The analytes are then either thermally desorbed from the fiber into the injector port of a gas chromatograph or removed by solvents for high performance liquid chromatography (HPLC) or electrophoresis applications. This technique has become popular for the analysis of organic compounds because it combines sampling and pre-concentration in a single step. The main advantages are faster analysis, low price of the fiber, fibers can be reused from several to thousands of times (depending on the extraction and desorption conditions), no solvents or complicated apparatus are required and good results are obtained over a wide range of analyte concentrations. However, SPME is not an exhaustive extraction technique – only 7–10% of the analyte present in the sample is extracted.

Stir bar sorptive extraction (SBSE) was introduced by Baltussen *et al.*, using stir bars 10 mm long for sample volumes of 10–50 mL and 40 mm long for sample volumes up to 250 mL, coated with 55 and 219 µL, respectively, of polydimethylsiloxane (PDMS) liquid phase [7]. However, in recent years other polymeric coatings such as polysiloxane [8] or polytetrafluoroethylene (PTFE) [9] have been proposed by several authors. In SBSE, the stir bar is typically immersed in the sample, which is stirred for a specific time, usually between 30 and 60 min depending on the sample volume and the stirring speed, to approach equilibrium between the sorbent and the matrix. The stir bar is then removed from the aqueous sample and the absorbed compounds are either thermally desorbed and analyzed by gas chromatography (GC), for very high sensitivity, or desorbed by means of a liquid, for improved selectivity or for interfacing to an HPLC system [10]. SBSE is not used as an exhaustive extraction procedure. However, since larger volumes of sorbent phase are used in the SBSE (50–200 µL) than in SPME (0.5 µL), recoveries from aqueous solutions are greater than recoveries obtained by SPME [3]. Thus, sensitivities 100–1000 times higher for SBSE than by SPME have been reported for the extraction of analytes from beverages [11]. However, in SBSE more time is required to reach equilibrium because more analyte mass is transferred to the sorbent phase [3]. This technique can be used for the extraction of organic compounds of a broad range of polarities from aqueous solutions (beverages, aqueous food, biological and environmental samples) with detection limits in the low to even sub-ppt range [7].

3.3
Ultrasonic Assisted Extraction (UAE)

3.3.1
Basic Principles

Extraction assisted by ultrasound is an advantageous alternative to both conventional (LLE, SPE, SLP and Soxhlet) and, occasionally, modern (e.g., SFE and MAE) extraction techniques. Ultrasonic assisted extraction is based on a sonochemical phenomenon associated with acoustic cavitation, that is, with the formation of microbubbles in a liquid when a large negative pressure is applied to it. These bubbles grow to an unstable size and subsequently collapse violently, releasing an intense local energy with important chemical and mechanical effects [12]. During bubble collapse instantaneous temperatures of several thousand degrees and pressures in excess of 1000 atmospheres are generated [12]. Moreover, the implosion of cavities establishes an unusual environment for chemical reactions [13]. Chapter 1 gives a detailed review of the power of ultrasound and its cavitation effects.

As a result of the cavitation process, intimate contact between the sample and the extraction solvent is ensured. Thus, ultrasonic radiation is a powerful tool to facilitate and accelerate different stages of the analytical process, for instance the extraction of organic compounds from liquid and solid samples.

UAE has the potential to fulfill the requirements of a sensitive and reliable method. Its major advantages are [3, 14]:

1. It provides an efficient contact between the solid matrix and the solvent, usually resulting in a greater extraction of analyte.
2. It reduces the extraction time and the volume of organic solvent required for an efficient extraction when compared with other extraction techniques such as Soxhlet or LLE.
3. It is possible to select the solvent or solvent mixture that allows the maximum extraction efficiency and selectivity.
4. A wide range of sample sizes can be used.
5. Several extractions can be performed simultaneously.
6. It is a highly reproducible and effective technique, and no specialized laboratory equipment is required. The equipment for ultrasonication is very simple, easy to operate and relatively cheap.

The principal drawback associated with UAE is sonochemical degradation. The chemical effects of ultrasound derive from acoustic cavitation. As a result of cavitation, extreme temperatures and pressures can be developed locally within the bubbles during their collapse; therefore, these bubbles operate as hot spot microreactors. Furthermore, it has been suggested that, during UAE, the solvent can dissociate and the resulting radicals can react with the analytes [15]. For instance, water may dissociate within the bubble to produce HO$^{\bullet}$ and H$^{\bullet}$ radicals that migrate towards the interface with the bulk solution. Therefore, it is very important to optimize the water content of the extraction solvent. The behavior of acids under an ultrasonic field is not yet well established [15]. Consequently, compounds can be degraded (i) directly via pyrolytic reactions occurring inside the bubble and/or at the interfacial region or (ii) indirectly via radical reactions occurring at the interface and/or in the solution bulk. Sonochemical degradation of some compounds, such as phenol and substituted phenols, chlorinated hydrocarbons, pesticides and PAHs, has been observed by several authors [16, 17].

3.3.2
Parameters Influencing Ultrasonic Assisted Extraction

Optimization of ultrasound operational parameters according to a specific type of analyte and matrix is of prime importance to achieve high extraction efficiency. Thus, the following parameters must be optimized: amount of sample, particle size, extraction solvent (type, pH and volume), sonic power applied, frequency, extraction time and extraction temperature. These factors and their effects in the extraction process are explained below.

3.3.2.1 Amount of Sample
To improve the detection limits of the method the amount of sample could be increased. However, different authors have observed that increasing the amount of sample, while retaining a constant volume of organic solvent, diminishes the extraction efficiency. This might be because the solvent volume is too low to extract the compounds efficiently [18–20]. Consequently, if higher amounts of samples are required, the volume of the extracting solvent should also be increased to keep constant the ratio of sample to solvent. In addition, it is important to keep in mind that an increment in sample size could also increase the matrix effects when mass spectrometry detection is used, as described by Gatidou and coworkers [19].

3.3.2.2 Sample Particle Size
This parameter should be optimized in the extraction of solid matrixes. In general, small particle sizes are recommended to increase the solid–liquid interface and thus to increase the extraction efficiency. If possible, samples may be ground to a fine powder before extraction. However, for volatile analytes, this step is not recommended because these compounds can be liberated during the grinding process.

3.3.2.3 Extraction Solvent
One of the most influential parameters governing the extraction efficiency and selectivity in UAE is the nature of the solvent. This is a critical parameter in multiresidue analysis where the extraction solvent should have polarity properties compatible with all analytes. The extraction solvent must be chosen taking into

account the polarity and solubility of the target compounds. In general, solvents with similar properties to the analyte provide good extraction efficiencies. Nevertheless, the presence of impurities in the matrix that can be co-extracted with the target compounds must be also considered. Thus, the solvent used should allow the effective extraction of target analytes without extract interfering compounds. For instance, high recoveries of triazine herbicides from mud samples from olive washing devices can be obtained using acetone as extraction solvent. Nevertheless, acetone also dissolve the cellulose filters, producing muddy extracts, more irregular background in the chromatograms and giving recoveries of about 200% for some compounds, mainly due to some interferents co-eluted. This problem can be avoided using mixtures of acetone/cyclohexane (1 : 3, v/v) [5]. In some cases, the presence of water can improve the extraction efficiency of analytes that show a hydrophilic–hydrophobic mixed character. For example, increments of about 50% in the recoveries of endocrine disruptors from soils have been obtained after the inclusion of 30% of water in the extraction solvent (70 mL of solvent + 30 mL of water) [21]. However, the required percentage of water in the medium is not well established yet. Thus, for some authors inclusion of 5% of water is enough [1], while others suggest the inclusion of 50% of water to the extracting solvent [22] for the extraction of pesticides from soil and sediments. It is important to keep in mind that the presence of water in the extraction solvent may produce sonochemical degradation effects [15], as explained above; because of this, the water content must be optimized.

3.3.2.4 pH of Extracting Solution
In some cases, it may be necessary to adjust the pH to ensure that compounds are in the appropriate form to achieve efficient desorption from the sample matrix. For instance, the use of potassium bicarbonate buffer at pH 9.7 as an extracting solvent is required for the extraction of mycophenolic acid from cheese [23].

3.3.2.5 Solvent Volume
As explained above, the volume of the organic solvent must be in concordance with the amount of sample to extract the compounds efficiently. The use of solvent volumes higher than the optimum does not improve recoveries. On the contrary, under these conditions the concentration of the analyte is diluted, causing at the same time a decrease in the sensitivity of the method and an unnecessary consumption of solvent.

3.3.2.6 Sonic Power
Different ultrasonic devices are available to speed up the liquid–liquid or solid–liquid extractions of organic compounds, namely, the ultrasonic bath and the ultrasonic probe. The former is more frequently used because the ultrasonic probe was considered to provide too much energy and, consequently, sonochemical degradation could be expected [24]. However, some applications with an ultrasonic probe have been described. For example, it has been used for the extraction of malic and tartaric acids from grapes and winemaking by-products [25] and also for the extraction of PCBs and OCPs from eggs [26].

3.3.2.7 Frequency

The frequency influences the formation of cavitation bubbles. At high frequencies the production of cavitation bubbles is more difficult, as explained in Chapter 1.

3.3.2.8 Extraction Time

The extraction time must be adjusted to obtain quantitative recoveries. Thus, extraction efficiency increases with the sonication time until equilibrium is reached. Extraction times of between 3 and 60 min are described in the literature. Longer extraction times could produce a significant reduction of the recoveries obtained, probably due to degradation of the compounds [19] or decomposition of the organic solvent caused by the effects of the ultrasound waves [27]. Thus, Melecchi and coworkers have observed a decrease in the yield mass of phytosterols, hydrocarbons and fatty acids methyl esters (FAMES) from flowers at extraction times longer than 300 min, which they ascribe to possible decomposition of the organic compounds by the effects of the sound waves [27]. A slight decrease in recovery values was also observed in the extraction of triazine herbicides after increasing the sonication time up to 40 min, probably because of decomposition of the herbicides due to heating [5]. In addition, at longer extraction times a reduction in chromatographic resolution may occur since the major compounds may overlap with minor ones [28]. Nevertheless, extraction times of 90 min have been applied to the extraction of plasticizers from packaged food without degradation [20]. To improve the recovery percentage, extraction cycles can be carried out [18, 21, 28].

3.3.2.9 Extraction Temperature

In general, extraction methods that use a high temperature, like Soxhlet, generally yield better recoveries than cold extraction procedures. For instance, some compounds such as pesticides are more soluble when the bonds are broken as a consequence of heat input. In UAE, a higher temperature means a higher efficiency in the extraction process due to the increase in the number of cavitation bubbles and in the surface contact area. However, this effect tends to disappear when the temperature is near the boiling point [25]. It must be borne in mind that the effect of temperature depends on the analyte. For some compounds, increasing the extraction temperature to 45–70 °C increases the recovery [21, 25, 29–31]. In contrast, other analytes can be easily degraded with an increment of temperature or, in the case of VOCs, with high vapor pressures, they can easily be lost by evaporation. When lower recoveries may be expected with an increment of temperature, this parameter must be controlled during the extraction step [28]. During ultrasonic extraction the solvent temperature increases with the extraction time and the sonic power applied, owing to the sonication process.

3.3.3 Applications

Ultrasonic assisted extraction has been proved to be an expeditious, inexpensive and efficient alternative to conventional extraction techniques and has been successfully applied to solid and liquid matrixes for the determination of different group of analytes, as it can be seen in several recent reviews [32–34]. Thus, UAE has been applied with success to the extraction from liquid and solid matrixes of different organics such as surfactants [13], aromas [28, 35–40], barbiturates [41], butyltin compounds [29, 42, 43], biologically active compounds [30], chlorophenols [44], endocrine disruptors [19, 21, 45], hydrocarbons (HC), fatty acids methyl esters (FAMES) and phytosterols [27], isoflavonoids [46], mycotoxins [23, 47, 48], oils [49], organic acids [25], plasticizers [14, 20], polycyclic aromatic hydrocarbons (PAHs) [50–52], polybrominated biphenyls (PBBs) and polybrominated diphenyl ethers (PBDEs) [53, 54], polychlorinated biphenyls (PCBs) [26, 31], pesticides [1, 5, 18, 22, 26, 55–64], solanesol [65, 66], soyasaponins [67], steroids and triterpenoids [68] and volatile organic compounds (VOCs) [4] among others. The principal parameters of the methods involving the extraction of the above-mentioned compounds, such as amount of sample, extraction solvent, sonication time, extraction temperature and parameters related to the analytical technique of measurement, are compiled in Table 3.1.

3.3.3.1 Liquid Samples

As it can be seen in Table 3.1, UAE is more frequently applied to solid matrixes than liquid ones. In general, extraction of volatile and semivolatile organics from liquid samples is easier than the extraction of the same compounds from solid matrixes because the mass transfer of analytes from the matrix to the extraction solvent is facilitated when they are dissolved in a liquid matrix.

With liquid matrixes, UAE has been applied to the extraction of aroma compounds from aged brandies, must, wine and honey samples [28, 35–40], and pesticides from honey samples [64]. Dichloromethane or mixtures of n-pentane/diethyl ether were the solvents preferentially used for the extraction of aroma compounds while mixtures of water/benzene were used in the case of pesticides. According to the results of Cocito et al. [69] and Peña et al. [28], an extractant volume equal to 30% of the sample volume should be enough to allow analyte extraction while avoiding the formation of stable emulsions. However other authors suggest the use of a sample to solvent ratio of 1 : 1 [38–40].

The addition of salt, depending on the salt type and its concentration, can improve the extraction efficiency of aroma compounds. As is well known, the addition of salt to aqueous media prompts a reduction in the solubility of analytes in the medium in which they are dissolved, improving the extraction process. Equilibrium depends mainly on the type and amount of salt added. Thus, NaCl and $MgSO_4$ have been used to improve the extraction of aromatic compounds from honey and wine samples [35–37, 39]. However, other authors achieved efficient extractions of these compounds without salt addition [28, 38, 40].

3.3.3.2 Solid Samples

For solid samples, extraction is more complicated because direct contact between the analyte and the extraction solvent is more difficult. Thus, extraction of analytes from solid matrixes normally involves two steps: desorption of the target compounds from

Table 3.1 Substances isolated using ultrasonic assisted extraction and the principal parameters of these methods.

Matrix	Analyte	Sample size	Solvent	T^a (°C)	Sonication time (min)	Clean-up	Recoveries (%) (±RDS %)	Ref.
Anionic surfactants	Soils	5 g	100 mL of methanol	RT	10	SPE, Isolute SAX + Isolute C18 cartridges	75 ± 2	[13]
Aroma compounds	Flowers	5 g	30 mL of n-pentane: diethyl ether (1:2, v/v)	25	10	—	±20	[35]
	Honey	40 g	30 mL of n-pentane: diethyl ether (1:2, v/v) in presence of MgSO$_4$	RT	10	—	±20	[35, 36]
	Wine	25 mL	10 mL of dichloromethane in presence of NaCl	25	15	—	±6	[37]
		50–100 mL	25–50 mL of dichloromethane	20	10 (×3)	—	93–98 ± 6	[28, 38]
		100 mL	50 mL of diethyl ether: n-pentane (2:1, v/v) in presence of MgSO$_4$	20	10 (×3)	—	±21	[39]
	Brandies	100 mL	50 mL of dichloromethane	20	10 (×3)	—	45–113 ± 18	[40]
Barbiturates	Pork	2 g	50 mL of acetonitrile in presence of Na$_2$SO$_4$	30	30 (×2)	SPE, MWCNTs-packed cartridges	75–96 ± ±8	[41]
Biologically active compounds	Plants	0.1 g	1 mL of methanol: ethanol (90:10, v/v)	50	3	SPE, C18 cartridges	99 ± 2	[30]

(Continued)

Table 3.1 (Continued)

Matrix	Analyte	Sample size	Ultrasonic conditions				Recoveries (%) (±RDS %)	Ref.
			Solvent	T^a (°C)	Sonication time (min)	Clean-up		
Butyltin compounds	Sediments	0.2 g	5 mL of glacial acetic acid	RT	4	—	70–90 ± 10	[42]
		0.1–0.5 g	5 mL of glacial acetic acid	RT	4	HS-SPME, PDMS-100 µm	±12	[43]
Chlorophenols	Soils	0.5–2 g	20 mL of acetic acid	50	30	—	94–101 ± 6	[29]
		6 g	48 mL of water	RT	30	SPME, C[4]/OH-TSO-40 µm	81–99 ± 7	[44]
Endocrine disruptors	Sewage sludge	0.02 g	8 mL of methanol : water (5 : 3, v/v)	50	30	—	53–116 ± 13	[19]
	Soils	2 g	5 mL of methanol : water (70 : 30, v/v)	45	15 (×2)	SPE, C18 cartridges	26–95 ± 15	[21]
	Sediments	5 g	First 7 mL of acetone, then 7 mL of methanol and finally 7 mL of dichloromethane	35	20 (×3)	—	59–85 ± 18	[45]
HC, FAMES, phytosterols	Flowers	5 g	150 mL of methanol	RT	140	SPE, silica gel column		[27]
Isoflavonoids	Plants	1 g	10 mL of methanol : water (90 : 10, v/v)	50	60	—		[46]
Mycotoxins	Cheese	0.5 g	5 mL of potassium bicarbonate buffer (0.2 M, pH 9.7)	RT	30	SPME, CW/TPR-100 µm	50–94 ± 5	[23]
		0.5 g	3 mL of methanol	RT	5	SPME, CW/TPR-100 µm	±5	[48]

Analyte	Matrix	Sample amount	Solvent	Temp.	Time (min)	Technique	Recovery (%)	Ref.
	Cornflake	0.5 g	5 mL of methanol/2% KHCO$_3$ (70:30, v/v)	RT	5	SPME, PDMS/DVB 60 μm	74–103 ± 9	[47]
Oil	Tobacco seeds	5 g	50 mL of hexane	Boiling T^a	20	—	78.2	[49]
Organic acids	Grapes	1 g	100 mL of water	70	30	—	90 ± 5	[25]
Plasticizers	Plastics	0.01–1.5 g	4–10 mL of methanol	RT	30	SPME, C[4]/OH-TSO-40 μm	95–102 ± 10	[14]
	Curry paste	3 g	20 mL of dichloromethane:cyclohexane (1:1, v/v)	RT	90	SPE, florisil cartridge	88–99 ± 9	[20]
PAHs	Aerosol collector filters	1 filter	30 mL of dichloromethane	RT	15 (×2)	SPE, C$_{60}$ silica gel column	95	[50]
	Bark	1.5 g	80 mL of dichloromethane	RT	20 (×4)	SPE, SiO$_2$ cartridges		[51]
	Sediments	10 g	100 mL of acetonitrile	RT	15 (×5)	LLP with a mixture of water/benzene	62–86 ± 15	[52]
	Mussels	1 g	9 mL of acetone:hexane (50:50, v/v)	RT	15 (×4)	SPE, silica gel column	70–84 ± 4	[52]
PBBs and PBDEs	Soils, sediments and sewage sludge	2 g	8 mL of hexane	RT	15	HS-SPME, PDMS-100 μm	92–138 ± 9	[53]
	Soil and dust	0.15 g	2 mL of acetone	RT	20	SBSE, PDMS-β-CD stir bar	56–118 ± 18	[54]
PCBs	Bird livers	1 g	20 mL of n-hexane:acetone:dichloromethane (3:1:1, v/v/v) in presence of Na$_2$SO$_4$	RT	15 (×2)	sulfuric acid treatment + HS-SPME, PDMS-100 μm	63–94 ± 11	[31]
	Eggs	2–5 g	30 mL of hexane	RT	5 (UP)	SPE, florisil columns	50–116 ± 29	[26]
Pesticides	Soils and sediments	5 g	5 mL of acetone:water (95:5, v/v)	RT	30	SPME, polyacrylate 85 μm	>91 ± 9	[1]
		5 g	40 mL of cyclohexane:acetone (3:1, v/v)	RT	20	—	95–107 ± 10	[5]

(Continued)

Table 3.1 (Continued)

Matrix	Analyte	Sample size	Ultrasonic conditions Solvent	T^a (°C)	Sonication time (min)	Clean-up	Recoveries (%) (±RDS %)	Ref.
		2 g	20 mL of methanol	50	30 (×2)	SPE, C18 cartridges	86–106 (±15)	[18]
		5 g	10 mL of acetone:water (1:1, v/v)	50	30	SPE, disks of C8	55–100 ± 13	[22]
		0.5 g	5 mL of methanol	RT	15	SPME, PDMS-100 μm	47–90 (±16)	[55]
		30 g	40 mL of methanol:water (4:1, v/v)	RT	20	SPE, Oasis HLB cartridges	60–98 ± 11	[56]
		2–4 g	30 mL of dichloromethane:acetone:ethyl acetate:cyclohexane (2:1:1:1 v/v/v/v)	RT	1 min with 3 s pulses (×3)	—	61–124 ± 16	[57]
		1 g	5 mL of dichloromethane	RT	20 (×2)	—	80–109 ± 14	[58]
		5 g	First 7 mL of acetone, then 7 mL of dichloromethane and finally 7 mL of hexane	RT	20 (×3)	SPE, SDB disks	71–84 ± 14	[59]
Bird livers		1 g	20 mL of n-hexane:acetone (4:1, v/v) in presence of Na_2SO_4	RT	30	Sulfuric acid treatment + HS-SPME, PDMS-100 μm	±14	[63]
Eggs		2–5 g	30 mL of hexane	RT	5 (UP)	SPE, florisil columns	51–108 ± 22	[26]
honey		5 g	60 mL of benzene:water (1:1, v/v)	35	20 (×3)	—	92–95 ± 3	[64]

Analyte	Matrix	Mass	Solvent	T^a	Time (min)	Post-treatment	Recovery (%)	Ref.
	Mud	5 g	40 mL of cyclohexane : acetone (3 : 1, v/v)	RT	20	SPE, alumina column	70–123	[56]
	Tea	2 g	20 mL of acetone : n-hexane (1 : 1, v/v)	RT	30	SPE, a cartridge with activated carbon + florisil	44–81	[62]
	Tobacco	0.2 g	25 mL of water	RT	10	SPME, PDMS-100 μm	80–96 ± 12	[60]
		5 g	50 mL of acetonitrile	RT	30	LLP, with dichloromethane	90–93 ± 3	[61]
Solanesol	Tobacco	2 g	60 mL of hexane	RT	20 (×3)	—	96–105 ± 4	[65]
		2 g	60 mL of methanol	25	20 (×3)	—	97–100 ± 3	[66]
Soyasaponins	Hypocotyls	0.1 g	75 mL of ethanol : water (40 : 60, v/v)	25	20 (×3)	—	99 ± 13	[67]
Steroids and triterpenoids	Plants	1 g	20 mL of hexane	30	30	SPE, sep-kap column		[68]
VOC	Clays	6–10 g	20 mL of methanol	45	30–60	—	100 ± 2	[4]

T^a: Temperature expressed in °C
RT: Room temperature

the solid matrix and then their solubilization into the extraction solvent. Ultrasound permits the disruption of cell walls, desegregation of the solid sample and particle-size reduction, as a result of cavitation bubble collapse, thereby improving mass transfer of analytes into the solvent. For this reason, ultrasound is a major advance for the extraction of organics from solid samples. Consequently, UAE has been frequently applied to environmental samples (clays, soils, sediments and sludge) [1, 4, 5, 13, 18, 19, 21, 22, 29, 42–45, 52–59], foods (cheese, cornflakes, eggs, meat and mussels) [23, 25, 26, 41, 47, 48, 52], plants [27, 30, 35, 46, 49, 60–62, 65–68] and biological material [31, 63].

To ensure optimum extraction efficiencies for organic compounds, solid samples such as soils and sediments should be ground to a fine powder before extraction. In the same way, it is recommended to finely grind biological samples with a drying agent such as sodium sulfate prior to submit the sample to the extraction process. Although the manipulation of the sample with sodium sulfate is relatively time-consuming, results reported have demonstrated that this step is crucial and must be included in the sample treatment to avoid appreciable losses of target compounds. Thus, sodium sulfate has been used in the extraction of barbiturates in pork [41] or PCBs and pesticides in bird livers [31, 63].

3.3.3.3 Clean-Up

Usually, the extracts must be cleaned-up before chromatographic determination because solvents are non-selective and, therefore, tend to extract endogenous material from the matrix, which produces spurious peaks on the chromatogram. Because of this, a clean-up step is generally needed to achieve satisfactory detection limits [22]. For this purpose the following extraction techniques can be considered: liquid–liquid partitioning (LLP) with dichloromethane or mixtures [52, 61]; SPE with cartridges or columns packed with different sorbents (C_8, C_{18}, florisil, alumina, Isolute SAX, Oasis HLB, SDB, silica gel, etc.) [13, 18, 20–22, 26, 27, 30, 41, 42, 50–52, 56, 59, 62, 68]; SPME using fibers of polydimethylsiloxane (PDMS), polyacrylate (PA), polydimethylsiloxane/divinylbenzene (PDMS/DVB), Carbowax/Templated Resin (CW/TRP) [1, 14, 23, 43, 44, 47, 48, 53, 55, 60]; or SBSE with stir bars of PDMS-β-CD [54].

The clean-up step is especially important for biological samples that contain large quantities of lipids. The extraction of trace compounds such as CPs and PCBs in the presence of major sample extractable components such as lipids poses special challenges in gas chromatography. Large amounts of fat may cause problems in the injection port of the gas chromatograph and at the top of the chromatographic column. In addition, when using mass spectrometry detection, the ion source might become contaminated, causing impaired analytical performance and signal suppression [31, 63]. The most effective clean-up method for removing lipids involves the use of concentrated sulfuric acid, which can remove more than 90% of the lipids present in the extract. However, sulfuric acid clean-up must be applied cautiously, because it completely or partially destroys many labile compounds [26]. Sulfuric acid treatment in combination with HS-SPME using PDMS fibers is the clean-up process suggested by Lambropoulou and coworkers for the determination of CPs and PCBs in bird livers [31, 63].

3.4
Coupling Ultrasound with Other Extraction Techniques

3.4.1
Coupling Solid Phase Microextraction (SPME) and Ultrasound

SPME consists of two steps. First, with the fiber kept in the microsyringe needle, the septum of the sample vial is perforated and then the fiber is exposed to the sample for a specific period of time. After equilibrium is reached, or at a specific time prior to achieving equilibrium, the fiber is retracted into the microsyringe needle and withdrawn from the sample. Second, the fiber is desorbed thermally into the injector port of a gas chromatograph or by solvents for HPLC or electrophoresis applications.

The basic principles and technical aspects of this technique have been summarized recently [70]. This review includes details of the optimization of the extraction process and also numerous applications of SPME in various fields of analytical chemistry. SPME coupled with GC has been applied to the analysis of pesticides, PCBs, PAHs, VOCs, phenols, nitrophenols, fatty acids and organometallic compounds, among others [70, 71].

To achieve good recoveries different parameters must be optimized, such as (i) sampling mode, (ii) selection of fiber coating, (iii) parameters that affect the adsorption of the analyte onto the fiber (agitation method, extraction temperature, sample volume, salt addition, pH adjustment and sampling time) and (iv) parameters that affect the desorption process (desorption time, desorption temperature or solvent selection).

The use of sonication in the SPME procedure has been reported by different authors with the following aims:

1. To improve the extraction procedure in direct-SPME [72–74].

2. To improve the extraction procedure in the headspace mode (HS-SPME), increasing the pass of volatile aromatic compounds to the headspace and reducing the extraction time required [75–78]; and

3. To facilitate desorption of compounds adsorbed/absorbed in the fiber into the extraction solvent [79, 80].

The following sections explain in detail the advantages of coupling ultrasonication and SPME.

3.4.1.1 Improving the Extraction Procedure in Direct-SPME
Normally, agitation is used to enhance the diffusion of analytes towards the fiber. Thus, an adequate agitation system can afford an important reduction in extraction time and lower standard deviations. Theoretically, the time required to reach equilibrium from well-agitated solutions is proportional to the thickness of the polymer coating and inversely proportional to the diffusion coefficient of the analyte in the coating. However, if the agitation is not efficient, the diffusion rate through the static aqueous layer adjacent to the fiber surface becomes the major factor in determining the time required to reach equilibrium.

Agitation can be produced by magnetic stirring, intrusive mixing or by sonication. In 1993, Motlagh and Pawliszyn studied the applicability of these three agitation systems for SPME applications [72]. They observed that magnetic stirring provided low mixing efficiency, despite being the most common agitation method used. Moreover, it is important to ensure that the rotation speed of the stir bar is constant and the base plate is thermally isolated from the vial containing the sample. Otherwise, heating of the sample might result in loss of precision by thermal degradation of the analytes [81]. With respect to intrusive mixing, this technique allows efficient agitation but causes significant sample heating with consequent loss of analyte. Sonication was found to be the best option for sample agitation, allowing high yields of analyte recovery in short extraction times. The authors also tested several sonication powers, ranging from 0 to 150 W. The results showed that low power sonication was enough to perform most extractions successfully, avoiding, in this manner, the drawbacks of high-power sonication such as sample heating, sample decomposition and high-pitched noise.

Boussahel et al. have used an ultrasonic bath at 50 °C to accelerate the extraction of chlorinated pesticides (lindane, heptachlor, 2,4-DDE, 4,4-DDE, α-endosulfan, β-endosulfan, 2,4-DTT and 4,4-DTT) from waters by direct-SPME with a PDMS-100 µm fiber [73]. This work suffers from two major drawbacks. Firstly, the use of high temperatures is normally inadvisable because SPME extraction is an exothermic process and, consequently, high temperatures make analyte transference from the matrix to the SPME fiber difficult. In addition, thermal degradation of the analytes could take place at high temperatures. However, with chlorinated pesticides, which are very stable, no thermal degradation is expected. Secondly, the authors do not report recoveries for these compounds for sample treatment without sonication. Thus, the extraction efficiency of the method could be attributed to the heating effect and not to the ultrasonic power.

More recently, Rial-Otero et al. have reported a new SPME method for the extraction of acaricides (amitraz, bromopropylate, coumaphos and fluvalinate) from honey [74]. In this work a comparison was made between the classic stirring method using a stir bar and the sonication method using an ultrasonic bath. Moreover, different sonication frequencies (35 and 130 kHz) and different working modes (sweep, standard and degas) were studied and optimized to speed up the microextraction procedure. Notably, ultrasonic baths with a dual frequency of ultrasonication, such as the used in this work, are relatively new in the analytical laboratory. Acaricide recoveries obtained using sonication were higher than those obtained with magnetic stirring. The use of low ultrasonic frequency, 35 kHz, provides worse results than the use of the high frequency, 130 kHz, especially in the case of amitraz. This can be related to the cavitation effects produced by the ultrasonic bath at 35 kHz, which would be enough to decompose amitraz in the acidic media. Cavitation effects are directly linked to the sonication frequency. The lower the frequency the higher the cavitation effects for the same amplitude [12]. It must be also stressed that low sonication frequencies are recommended for solid–liquid extraction of analytes. Therefore, the use of low frequencies could result in an increment of analyte desorption from the fiber, and therefore in a decrease in extraction efficiency. The best extraction results were achieved with the polyacrylate fiber, performing the extraction for 30 min in an ultrasonic bath operating at 130 kHz in the standard mode.

3.4.1.2 Improving the Extraction Procedure in HS-SPME

In the headspace, diffusion of analytes to the fiber is very fast and sample agitation does not improve the diffusion process. However, agitation facilitates the equilibrium between the headspace and the aqueous phase during SPME sampling, thus reducing the depletion of the headspace concentration [81]. Sonication is the most efficient agitation method evaluated to date for SPME applications [82]. Moreover, sonication can break down the structure of a sample matrix and release the volatile compounds physically trapped inside into the headspace. For example, Lee et al. have studied the influence of different sonication times (0–60 min) on the headspace concentration of volatile compounds of Parmesan cheese [75]. Total peak areas increased by 70% up to 40 min of sonication, and then decreased significantly after 40 min. In further work, the same authors made a comparative study of the effects of magnetic stirring and sonication on the extraction of volatile compounds from Kimchi (a traditional Korean fermented vegetable product) by HS-SPME [76]. An increment of about 16% in method sensitivity was observed with the use of an ultrasonic bath at 40 °C when compared with values obtained under the effect of heating only. However, better results were obtained with magnetic stirring. In this case, an increase of about 68% in method sensitivity was observed.

Kusch and Knupp have employed an ultrasonic bath to speed up the extraction of residual styrene monomer and other organic compounds in expanded polystyrene by HS-SPME [77]. They also tested the effect on the extraction procedure of different temperatures (25, 40, 60 and 80 °C). Good results, in terms of repeatability, were obtained after 15 min of sonication at 60 °C. As stated by Pawliszyn, heating a sample to an elevated temperature provides energy for analyte molecules to overcome the energy barriers that bind them to the matrix, enhances the mass transfer process and increases the vapor pressure of the analyte [82]. Nevertheless, application of high temperatures can adversely decrease their partition coefficients and subsequent extraction from headspace to the fiber, owing to the exothermic nature of the adsorption. To overcome this limitation, in 2004, Chia and coworkers developed a new device for HS-SPME screening of dibenzo-p-dioxins (PCDDs) and dibenzofurans (PCDFs) in heavily contaminated soil samples [78]. To improve evaporation of the analytes from the soil to the headspace a heater/ultrasonic screener was used with heated water at 85 °C. To achieve improved sensitivity, alcohol, chilled by circulating it through a refrigerated bath, was used to cool the fiber coating exposed to the headspace of the vial for 60 min to extract the target compounds. Using this sampling strategy, the mass transfer is accelerated and a temperature gap is created between the cold-fiber coating and the hot headspace, which significantly increases the distribution coefficients. Figure 3.3 shows a schematic diagram of this extraction device.

3.4.1.3 Facilitating the Desorption Process

As stated above, analytes adsorbed in the SPME fiber can be desorbed thermally into the injector port of a gas chromatograph or into a solvent for subsequent HPLC or

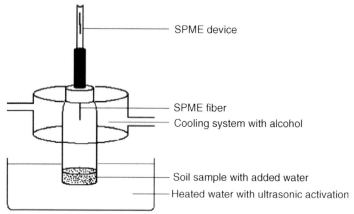

Figure 3.3 Schematic diagram of the extraction device developed by Chia and coworkers for HS-SPME.

electrophoresis analysis. In the last case, correct selection of the appropriate solvent and desorption time is crucial. To improve analyte desorption into the solvent, magnetic stirring or sonication can be used. Thus, Batlle and coworkers have used the power of an ultrasonic bath operating at 48 kHz to accelerate desorption of aliphatic isocyanates derivatized from the fiber into a mixture of acetonitrile/water (85 : 15, v/v) [79]. In addition, Yang et al. [80] have employed an ultrasonic bath to desorb chlorophyll from a chitosan membrane, used as SPME membrane, into NaOH (5%). Desorption times of 30 min were used.

3.4.2
Coupling Stir Bar Sorptive Extraction (SBSE) and Ultrasound

SBSE consists basically of two steps (Section 3.2): in the first, the stir bar is immersed into the sample, which is stirred for a specific time until equilibrium is reached; and in the second step the absorbed compounds are then either thermally desorbed into the injector port of the gas chromatograph or desorbed by means of a solvent for HPLC applications.

Ultrasonic treatment in the SBSE process has been implemented with two main aims: (i) to accelerate analyte desorption from the stirrers [8, 9, 83] and (ii) to avoid carryover [83]. Thus, Popp and coworkers have proposed the use of an ultrasonic bath, operating at 35 kHz, to facilitate the desorption of PAHs from stir bars of PDMS [83] or silicone (polysiloxane) rods [9]. Desorption times between 5 and 20 min were evaluated. After 10 min sonication of the stirrers in 100–150 µL of acetonitrile or an acetonitrile/water mixture (4 : 1, v/v) recoveries ranged between 58 and 100%. Results obtained for the silicone rods were comparable with those achieved with the PDMS stir bars [83]. The only disadvantage is that the procedure is not fully automated and that the extraction time is longer than when using SBSE (3 h using silicone rods vs. 1 h using SBSE). Sulistyorini et al. have used the power of ultrasound to desorb phenanthrene absorbed into the surface of PTFE stir bars [8]. In this case,

PTFE stir bars were placed in a glass vial with 3 mL of acetonitrile and sonicated for 15 min.

The effect of the type of vial used in the analyte desorption assisted by ultrasound has also been evaluated by Popp et al. [83]. For correct desorption, ultrasonic energy must be efficiently transferred from the sonication source to the solvent containing the stir bar. Thus, three different vials were used. The first was a glass screw top vial with a conical interior (250 µL). The second was a 2 mL vial with a glass flat-bottom insert (250 µL), which was used in two different modes: (i) the insert was placed into the vial as usual and (ii) the space between the insert and the inner wall of the vial was filled with water to improve energy transfer from the sonication source to the solvent. The results obtained revealed no significant differences among the three types of vials used. In addition, the authors observed that analyte carryover processes, from one sample to another, can be avoided by sonicating the stir bars in an appropriate solvent [83]. Thus, in the extraction of PAHs from water samples, the authors suggest that stir bars should be sonicated for 5 min in a 1-mL solution of methylene chloride and methanol (1 : 1, v/v). In addition, this purification should be realized immediately before the next sample extraction, to avoid carryover processes and new contamination during the time elapsed between purification and reuse of the stirrer.

3.5
Comparison between UAE and Other Extraction Techniques

Results reported in the literature comparing UAE with other extraction techniques are contradictory. On the one hand, results obtained by UAE were comparable or superior to those obtained by other techniques. For instance, Schinor et al. have observed that for the determination of steroids and triterpenoids in crude extracts of plants the mass yields obtained by UAE after 30 min of sample treatment are comparable to those achieved by maceration extraction for 24 h [68]. This means that the UAE was 48 times faster than the conventional method. In addition, Alissandrakis and coworkers have investigated and evaluated different isolation techniques [hydrodistillation (HD), microsimultaneous steam distillation-solvent extraction (MSDE), PT, SPME and UAE] for honey aroma compounds, observing great variability in the aroma compounds obtained, depending on the procedure employed [35, 36]. The authors conclude that the drastic HD and MSDE conditions lead to the formation of artifacts and also destroy sensitive compounds. However, UAE does not require heat, thereby avoiding the thermal generation of artifacts. Moreover, low and high molecular weight compounds can be extracted with this technique, providing good potential markers for honey origin control.

In contrast, for other types of analytes, lower extraction efficiencies have been obtained by UAE. For example, a much lower extraction efficiency was obtained by UAE than with matrix solid phase dispersion (MSPD) for the extraction of isoflavonoids in dried roots [46], especially for the major components of formononetin and calycosin; the efficiency with UAE was also lower than that obtained by Soxhlet. In addition, in the extraction of different organochlorine pesticides and pyrethroids from

Chinese teas, Ji *et al.* have achieved better results, in terms of extraction efficiency and clean extracts, using microwave assisted steam distillations than with UAE [62].

3.6
Conclusion

Ultrasound assisted extraction (UAE) has demonstrated good performance for the extraction of organics from liquid and solid samples. Satisfactory results are achieved faster than when using more traditional techniques. Moreover, small volumes of organic solvents are used, lowering the cost and shortening even more the time of analysis, because the samples can be subjected directly to the clean up step. Thus, rotary evaporation, which is normally employed when large volumes of organic solvents are used for extraction and often leads to losses of the most volatile compounds, can be avoided. In addition, due to the low costs of ultrasonic baths, UAE is as a viable procedure for many laboratories. Despite these desirable aspects, however, caution is necessary to avoid sonochemical degradation during ultrasound irradiation.

Abbreviations

ASE	Accelerated solvent extraction
CPs	Chlorinated pesticides
CW/TRP	Carbowax/templated resin
FAMES	fatty acids methyl esters
GC	Gas chromatography
HC	Hydrocarbons
HD	Hydrodistillation
HPLC	High performance liquid chromatography
HS	Static headspace analysis
LLE	Liquid–liquid extraction
LLP	Liquid–liquid partitioning
MAE	Microwave assisted extraction
MSDE	Microsimultaneous steam distillation–solvent extraction
MSPD	Matrix solid phase dispersion
OCPs	Organochlorine pesticides
OPPs	Organophosphorus pesticides
PA	Polyacrylate
PAHs	Polycyclic aromatic hydrocarbons
PBBs	Polybrominated biphenyls
PBDEs	Polybrominated diphenyl ethers
PCBs	Polychlorinated biphenyls
PCDDs	Dibenzo-*p*-dioxins
PCDFs	Dibenzofurans
PDMS	Polydimethylsiloxane

PDMS/DVB	Polydimethylsiloxane/divinylbenzene
PLE	Pressurized liquid extraction
PT	Dynamic headspace analysis or purge-and-trap
PTFE	Polytetrafluoroethylene
RT	Room temperature
SBSE	Stir bar sorptive extraction
SFE	Supercritical fluid extraction
SLP	Solid–liquid partitioning
SPE	Solid phase extraction
SPME	Solid phase microextraction
UAE	Ultrasonic assisted extraction
USE	Ultrasonic solvent extraction
USEPA	United States Environmental Protection Agency
VOCs	Volatile organic compounds

References

1 Lambropoulou, D.A. and Albanis, T.A. (2004) *Analytica Chimica Acta*, **514**, 125.
2 USEPA: Test methods for evaluating solid waste, physical/chemical methods SW-846. http://www.epa.gov/epawaste/hazard/testmethods/sw846/online/index.htm (last access in September 2008).
3 Winefordner, J.D. and Mitra, S. (eds), (2003) *Sample Preparation Techniques in Analytical Chemistry*, Chemical Analysis: A Series of Monographs on Analytical Chemistry and its Applications, Vol. 162, John Wiley & Sons, Inc., New Jersey, USA.
4 Dincutoiu, I., Górecki, T. and Parker, B.L. (2003) *Environmental Science & Technology*, **37**, 3978.
5 Guardia Rubio, M., Banegas Font, V., Molina Díaz, A. and Ayora Cañada, M.J. (2006) *Analytical Letters*, **39**, 835.
6 Arthur, C.L. and Pawliszyn, J. (1990) *Analytical Chemistry*, **62**, 2145.
7 Baltussen, E., Sandra, P., David, F. and Cramers, C. (1999) *Journal of Microcolumn Separations*, **11**, 737.
8 Sulistyorini, D., Burford, M.D., Weston, D.J. and Arrigan, D.W.M. (2002) *Analytical Letters*, **35**, 1429.
9 Popp, P., Bauer, C., Paschke, A. and Montero, L. (2004) *Analytica Chimica Acta*, **504**, 307.
10 Baltussen, E., Cramers, C.A. and Sandra, P.J.F. (2002) *Analytical and Bioanalytical Chemistry*, **373**, 3.
11 Hoffman, Bremer, R., Sandra, P. and David, F. (2000) *Labor Praxis*, **24**, 60.
12 Mason, T.J. (1999) *Sonochemistry*, Oxford University Press Inc., New York.
13 Nimer, M., Ballesteros, O., Navalón, A. et al. (2007) *Analytical and Bioanalytical Chemistry*, **387**, 2175.
14 Li, X., Zeng, Z., Chen, Y. and Xu, Y. (2004) *Talanta*, **63**, 1013.
15 Filgueiras, A.V., Capelo, J.L., Lavilla, I. and Bendicho, C. (2000) *Talanta*, **53**, 433.
16 Psillakis, E., Ntelekos, A., Mantzavinos, D. et al. (2003) *Journal of Environmental Monitoring*, **5**, 135.
17 Nakui, H., Okitsu, K., Maedab, Y. and Nishimura, R. (2007) *Journal of Hazardous Materials*, **146**, 636.
18 Gatidou, G., Kotrikla, A., Thomaidis, N.S. and Lekkas, T.D. (2004) *Analytica Chimica Acta*, **505**, 153.
19 Gatidou, G., Thomaidis, N.S., Stasinakis, A.S. and Lekkas, T.D. (2007) *Journal of Chromatography. A*, **1138**, 32.

20 Kueseng, P., Thavarungkul, P. and Kanatharana, P. (2007) *Journal of Environmental Science and Health Part B-Pesticides Food Contaminants and Agricultural Wastes*, **42**, 569.

21 Núñez, L., Turiel, E. and Tadeo, J.L. (2007) *Journal of Chromatography. A*, **1146**, 157.

22 Redondo, M.J., Ruiz, M.J., Boluda, R. and Font, G. (1996) *Journal of Chromatography. A*, **719**, 69.

23 Zambonin, C.G., Monaci, L. and Aresta, A. (2002) *Food Chemistry*, **78**, 249.

24 Santos, H.M. and Capelo, J.L. (2007) *Talanta*, **73**, 589.

25 Palma, M. and Barroso, C.G. (2002) *Analytica Chimica Acta*, **458**, 119.

26 Chu, S., Hong, C.S., Rattner, B.A. and McGowan, P.C. (2003) *Analytical Chemistry*, **75**, 1058.

27 Melecchi, M.I.S., Péres, V.F., Dariva, C. et al. (2006) *Ultrasonics Sonochemistry*, **13**, 242.

28 Peña, R.M., Barciela, J., Herrero, C. and García-Martín, S. (2005) *Talanta*, **67**, 129.

29 Nemanič, T.M., Milačič, R. and Ščančar, J. (2007) *International Journal of Environmental Analytical Chemistry*, **87**, 615.

30 Yang, Q., Zhang, X., Li, X. et al. (2007) *Analytica Chimica Acta*, **589**, 231.

31 Lambropoulou, D.A., Konstantinou, I.K. and Albanis, T.A. (2006) *Journal of Chromatography. A*, **1124**, 97.

32 Capelo, J.L. and Mota, A.M. (2005) *Current Analytical Chemistry*, **1**, 193.

33 Santos Júnior, D., Krug, F.J., De Godoi Pereira, M. and Korn, M. (2006) *Applied Spectroscopy Reviews*, **41**, 305.

34 Luque de Castro, M.D. and Priego-Capote, F. (2007) *Talanta*, **72**, 321.

35 Alissandrakis, E., Daferera, D., Tarantilis, P.A. et al. (2003) *Food Chemistry*, **82**, 575.

36 Alissandrakis, E., Tarantilis, P.A., Harizanis, P.C. and Polissiou, M. (2005) *Journal of the Science of Food and Agriculture*, **85**, 91.

37 Cabredo Pinillos, S., Cedrón Fernández, T., Gonzáles Briongos, M. et al. (2006) *Talanta*, **69**, 1123.

38 Gómez Plaza, E., Gil-Muñoz, R., Carreño Espín, J., Fernández López, J.A. and Martínez Cutillas, A. (1999) *European Food Research and Technology*, **209**, 257.

39 Hernanz Vila, D., Heredia Mira, F.J., Beltran Lucena, R. and Fernández Recamales, M. A. (1999) *Talanta*, **50**, 413.

40 Caldeira, I., Pereira, R., Clímaco, M.C., Belchior, A.P. and Bruno de Sousa, R. (2004) *Analytica Chimica Acta*, **513**, 125.

41 Zhao, H., Wang, L., Qiu, Y. et al. (2007) *Analytica Chimica Acta*, **586**, 399.

42 Carpinteiro, J., Rodríguez, I. and Cela, R. (2001) *Fresenius' Journal of Analytical Chemistry*, **370**, 872.

43 Carpinteiro, J., Rodríguez, I. and Cela, R. (2004) *Analytical and Bioanalytical Chemistry*, **380**, 853.

44 Li, X., Zeng, Z. and Zhou, J. (2004) *Analytica Chimica Acta*, **509**, 27.

45 Boti, V.I., Sakkas, V.A. and Albanis, T.A. (2007) *Journal of Chromatography. A*, **1146**, 139.

46 Xiao, H.B., Krucker, M., Albert, K. and Liang, X.M. (2004) *Journal of Chromatography. A*, **1032**, 117.

47 Aresta, A., Cioffi, N., Palmisano, F. and Zambonin, C.G. (2003) *Journal of Agricultural and Food Chemistry*, **51**, 5232.

48 Zambonin, C.G., Monaci, L. and Aresta, A. (2001) *Food Chemistry*, **75**, 249.

49 Stanisavljević, T., Lazić, M.L. and Veljković, V.B. (2007) *Ultrasonics Sonochemistry*, **14**, 646.

50 Borrás, E. and Tortajada-Genaro, L.A. (2007) *Analytica Chimica Acta*, **583**, 266.

51 Netto, A.D.P., Barreto, R.P., Moreira, J.C. and Arbilla, G. (2007) *Journal of Hazardous Materials*, **142**, 389.

52 Filipkowska, R., Lubecki, L. and Kowalewska, G. (2005) *Analytica Chimica Acta*, **547**, 243.

53 Salgado-Petinal, R., Llompart, M., García-Jares, C. et al. (2006) *Journal of Chromatography. A*, **1124**, 139.

54 Yu, R. and Hu, B. (2007) *Journal of Chromatography. A*, **1160**, 71.

55 Bouaid, R., Ramos, L., Gonzalez, M.J. et al. (2001) *Journal of Chromatography. A*, **939**, 13.

56 Belmonte Vega, R., Garrido Frenich, A. and Martínez Vidal, J.L. (2005) *Analytica Chimica Acta*, **538**, 117.

57 Lyytikäinen, M., Kukkonen, J.V.K. and Lydy, M.J. (2003) *Archives of Environmental Contamination and Toxicology*, **44**, 437.

58 Vagi, M.C., Petsas, A.S., Kostopoulou, M.N. et al. (2007) *Desalination*, **210**, 146.

59 Valsamaki, V.I., Sacas, V.A. and Albanis, T.A. (2007) *Journal of Separation Science*, **30**, 1936.

60 Shen, J., Xu, Z., Cai, J. and Shao, X. (2006) *Analytical Sciences*, **22**, 241.

61 Zhang, X., Cheng, X., Wang, C. et al. (2007) *Annali di Chimica-Rome*, **97**, 295.

62 Ji, J., Deng, C., Zhang, H. et al. (2007) *Talanta*, **71**, 1068.

63 Lambropoulou, D.A., Konstantinou, I.K. and Albanis, T.A. (2006) *Analytica Chimica Acta*, **573–574**, 223.

64 Rezić, R., Horvat, A.J.M., Babić, S. and Kaštelan-Macan, M. (2005) *Ultrasonics Sonochemistry*, **12**, 477.

65 Chen, J., Liu, X., Xu, X. et al. (2007) *Journal of Pharmaceutical and Biomedical Analysis*, **43**, 879.

66 Zhao, C.J., Li, C. and Zu, Y.G. (2007) *Journal of Pharmaceutical and Biomedical Analysis*, **44**, 35.

67 Yang, Y., Jin, M., Huang, M. et al. (2007) *Chromatographia*, **65**, 555.

68 Schinor, E.C., Salvador, M.J., Turatti, I.C.C. et al. (2004) *Ultrasonics Sonochemistry*, **11**, 415.

69 Cocito, C., Gaetano, G. and Delfini, C. (1995) *Food Chemistry*, **52**, 311.

70 Vas, G. and Vekey, K. (2004) *Journal of Mass Spectrometry*, **39**, 233.

71 Ghiasvand, R., Hosseinzadeh, S. and Pawliszyn, J. (2006) *Journal of Chromatography. A*, **1124**, 35.

72 Motlagh, S. and Pawliszyn, J. (1993) *Analytica Chimica Acta*, **284**, 265.

73 Boussahel, R., Bouland, S., Moussaoui, K.M. et al. (2002) *Water Research*, **36**, 1909.

74 Rial-Otero, R., Gaspar, E.M., Moura, I. and Capelo, J.L. (2007) *Talanta*, **71**, 1906.

75 Lee, J.H., Diono, R., Kim, G.Y. and Min, D.B. (2003) *Journal of Agricultural and Food Chemistry*, **51**, 1136.

76 Lee, J.H., Kang, J.H. and Min, D.B. (2003) *Journal of Food Science*, **68**, 844.

77 Kusch, P. and Knupp, G. (2002) *Journal of Separation Science*, **25**, 539.

78 Chia, K.J., Lee, T.Y. and Huang, S.D. (2004) *Analytica Chimica Acta*, **527**, 157.

79 Batlle, R., Colmsjo, A. and Nilsson, U. (2001) *Fresenius' Journal of Analytical Chemistry*, **371**, 514.

80 Yang, H.L., Wang, Y.R., Zhuang, Z.X. and Wang, X.R. (2006) *Spectroscopy and Spectral Analysis*, **26**, 11.

81 Eisert, R. and Pawliszyn, J. (1997) *Critical Reviews in Analytical Chemistry*, **27**, 103.

82 Pawliszyn, J. (1997) *Solid-Phase Microextraction: Theory and Practice*, Wiley-VCH Verlag GmbH, Weinheim.

83 Popp, P., Bauer, C. and Wennrich, L. (2001) *Analytica Chimica Acta*, **436**, 1.

4
Electrochemical Applications of Power Ultrasound
Neil Vaughan Rees and Richard Guy Compton

4.1
Introduction

When ultrasound is applied to a liquid, several processes can occur that often create unusual physical conditions: acoustic streaming, turbulent convection, microstreaming in the presence of oscillating bubbles and cavitation [1, 2].

Acoustic streaming, which is a nonlinear effect, arises from the absorption of momentum from the ultrasound source by the liquid media. A resulting turbulent flow in the direction of the applied sonic field has been claimed to reach local flow rates of >10 m s^{-1} [3, 4]. Figure 4.1 shows a schematic diagram of the flow pattern due to a sonic horn placed perpendicular to a planar surface.

Ultrasound provides greatly enhanced mass transport for electrochemical experiments and according to Moriguchi [5] can be rationalized in terms of a thinned Nernst diffusion layer. The diffusion layer model (Figure 4.2) enables a simplified description of mass transport close to the electrode surface by assuming a stagnant sublayer at the surface and a linear concentration gradient across it.

Equation 4.1 can be used to describe transport to an insonated electrode for a uniformly-accessible electrode [4, 6]:

$$I_{\text{lim}} = \frac{nFADC_\infty}{\delta} \tag{4.1}$$

where I_{lim} is the limiting current, n is the number of electrons transferred, F the Faraday constant, D is the diffusion coefficient, A the electrode area, C_∞ the bulk concentration of the electroactive species and δ is the diffusion layer thickness. For aqueous media where the sonic horn is perpendicular to the electrode surface, it has been found that [7]:

$$I_{\text{lim}} \propto AD^{2/3}C_\infty \tag{4.2}$$

and so:

$$\delta \propto D^{1/3} \tag{4.3}$$

Ultrasound in Chemistry: Analytical Applications. Edited by José-Luis Capelo-Martínez
Copyright © 2008 WILEY-VCH Verlag GmbH & Co. KGaA, Weinheim
ISBN: 978-3-527-31934-3

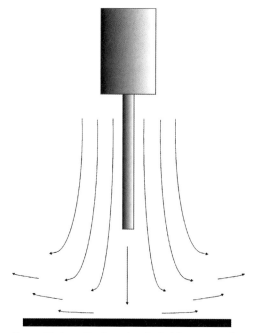

Figure 4.1 Acoustic streaming: typical flow pattern from an ultrasonic horn.

which suggests that the diffusion layer is not purely stagnant as suggested by the Nernst model, but contains some convective component akin to a hydrodynamic electrode [8]. Marken has shown that the minimum diffusion layer achievable using ultrasound in water is $0.7 \pm 0.1\,\mu m$ [4], which confirms that the use of power ultrasound can provide sufficiently high rates of mass transport to confer macroelectrodes with a similar kinetic timescale as microelectrodes.

Cavitation arises from the rapid compression and rarefaction phases of the longitudinal sound wave propagating through the liquid. The pressure variations

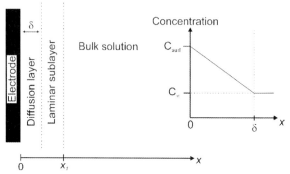

Figure 4.2 Schematic diagram of diffusion and boundary layers as applied in the diffusion layer model.

(a) Asymmetric bubble collapse

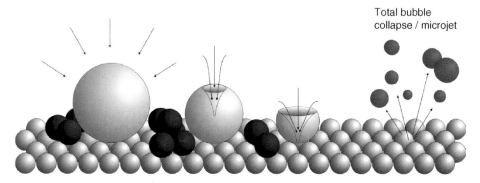

(b) Symmetric bubble collapse

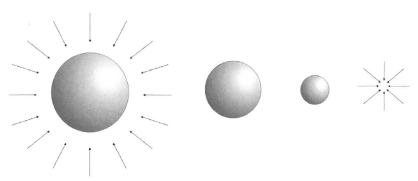

Figure 4.3 Cavitational collapse (a) at a surface and (b) in bulk solution. Adapted from [23].

can form stable or transient bubbles or voids within the liquid structure [9]. Noltingk and Neppiras [10] characterized a particular type of cavitation ("transient cavitation") where small bubbles in strong acoustic fields undergo an isothermal growth phase to several times their original size and then rapidly collapse adiabatically (Figure 4.3). This should be contrasted to "stable cavitation" where the bubble pulsates about an equilibrium size over successive acoustic cycles. Transient cavitation can produce several localized effects, including gaseous hot spots and gas shocks, which can both produce free radicals. The so-called thermal or "hot-spot" theory [2, 10, 11] contends that this cavitation can produce localized, transient high temperatures and pressures (c. 5000 K and 1700 bar, respectively) with low bulk liquid temperatures. Suslick has reported that clouds of collapsing cavitation bubbles can produce equivalent temperatures of 5000 K, pressures in excess of 1000 atm and heating/cooling rates above $1000 \, K \, s^{-1}$ [1, 11]. Whilst these values are estimates, other experimental work has showed they are not unreasonable [12–18]).

Cavitation occurs more readily at the solid–liquid interface, where surface imperfections enable vapor bubbles to easily form, than in bulk solution. When transient cavitation occurs near a surface, collapse becomes asymmetric driving a high-speed jet into the surface, termed "microjetting" (Figure 4.3) [19, 20]. This leads to surface cleaning, ablation and, sometimes, fracture. In the case of liquid–liquid interfaces, the result is efficient acoustic emulsification. Although this is an established mechanism, recent work has questioned how important a contribution it makes to sonovoltammetry at 20 kHz, indicating that bubbles are mostly close to their maximum size [21–23].

Klima [24–26], and Birkin and Silva-Martinez [9, 27, 28] have claimed, perhaps surprisingly, that the cavitational response at an electrode can be modeled as a transient response similar to that of a wall-tube electrode! The mechanism responsible for this behavior was thought to be imploding cavitation bubbles compressing the diffusion layer [9].

Ultrasound has several been reported to have effects on *chemical* processes, for example in homogeneous sonochemistry where novel species, or enhanced yields, and/or better selectivities can occur. However, in terms of *electrochemical* processes ultrasound appears to simply increase the rate of mass-transport. Whilst Compton reported no effect on the heterogeneous rate constant for electron transfer in an EC[1]) mechanism [29], Madigan and Coury claimed to see an increased rate whilst insonating in a suspension of alumina particles which had apparently raised the temperature of the solution at the electrode surface by 100 K [30]. Huck quantified the typical sonovoltammetric response for the ferricyanide redox, and found the results were generally consistent with those of rotating disk voltammetry [31]. Birkin has used sampled voltammetry at insonated microelectrodes with single and double potential step chronoamperometry to probe the electron transfer rates for the Ru$(NH_3)_6^{3+/2+}$ and $IrCl_6^{3-/2-}$ redox couples and found good agreement with results obtained under silent conditions [28]. In addition, Fontana has reported that for the $Fe^{2+/3+}$ and $Fe(CN)_6^{3-/4-}$ systems ultrasound has no other effect than to improve mass transport [32]. The conclusion is drawn, therefore, that ultrasound has little or no effect in simple electron transfer processes.

Sonovoltammetry has also been shown to be effective in probing homogeneous kinetics coupled to electrode processes. ECE[1])-type reactions have been widely used, for example the dehalogenation of halobenzophenones and nitrobenzenes [33, 34], and results found generally agree with those derived from silent, hydrodynamic or cyclic voltammetric studies [35, 36].

The ultrasonically-enhanced mass transport is thought to be due to two transient processes:

1. Bubble collapse at or near the solid–liquid interface with microjetting directed towards the electrode surface.
2. Bubble motion near or within the diffusion layer of the electrode.

While these processes aid mass transport, acoustic streaming is thought to be the major process of convective flux to the electrode surface [4, 6]. With an increase in

---

1) Reaction mechanisms in an electrochemical context are usually described using the notation introduced by Testa and Reinmuth [Testa, A.C. and Reinmuth, W.H. (1961) *Anal. Chem.* **33**, 1320]. In this convention, the letter E is used to denote a heterogeneous electron transfer step, and the letter C to indicate a homogeneous chemical step.

the local acoustic pressure, for example, it appears that there is a threshold for the detection of violent cavitational activity [37], perhaps due to a cushioning effect from the bubble populations. This threshold may coincide with the breakdown of the diffusion layer thickness model, even when calibration of the diffusion layer thickness is employed [34], such that a transition to a more complex mass transport model would be anticipated.

The depassivating effect of ultrasound depends on cavitational collapse at the solid–liquid interface [38]. Depending on the acoustic frequency and intensity, and the solvent used and temperature, the energy released can be considerable and damage to the electrode surface is very possible [7]. Evidence exists for the insonation of polished platinum (Figure 4.4) and glassy carbon electrodes [7, 39]. Surface images

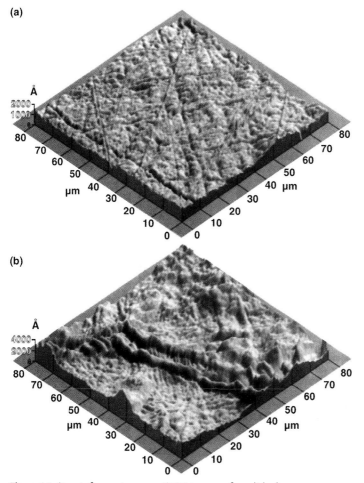

Figure 4.4 Atomic force microscopy(AFM) images of a polished polycrystalline Pt electrode (a) before and (b) after exposure to ultrasound for 120 s. Reproduced from [7] with permission from Elsevier.

taken over a period of a few minutes' insonation show pitting of the surface [7], the degree of which depends on the applied ultrasound intensity. This effect is exemplified by the oxidation of a nickel electrode surface in alkaline aqueous solution [40], and depassivation of iron electrodes [41]. Surface roughening/depassivation is of importance in sonoelectrochemistry, where normal electrochemical detection via anodic stripping voltammetry in highly passivating media would fail, as in some cases insonation can transform invisible signatures of metal species under investigation into large and quantifiable signals.

The formation of free radicals and other highly-reactive species within the gas bubble often results from transient collapse. For example, in water, hydroxyl radicals are formed by dissociation of water inside the cavitational bubble. Simple tests for radicals include the Weissler [42, 43] and Fricke [44] test solutions, and terphthalate [45, 46] and nitrophenol [47, 48] dosimeters. Evidence for the degradation products induced by ultrasound has been investigated by Reisz [49–52] using ESR spin-trapping of non-volatile nitrone, Hydroxyl radicals were verified as being generated in water by ultrasound by experiments in which radical scavengers compete with the spin traps for the OH and H radicals. Radicals have been found to be frequency dependent, for example the yield of hydrogen peroxide from the recombination of OH radicals is higher at 514 kHz than at 20 kHz [48]. Hydroxyl radicals have been found to be responsible for chemical effects such as the degradation of chlorophenols [53, 54] and cyanide [55] and the treatment of wastewater [56].

Electrosynthesis can also benefit from ultrasound. There are several examples [11, 57] where the increased mass transport accentuates the Faradaic processes compared to other background currents (Figure 4.5).

Ultrasound is also a powerful method of creating emulsions in immiscible systems, due to transient collapse of the liquid–liquid interface. Acoustic emulsions avoid the need for stabilizing reagents or surfactants. Furthermore, it enables the use

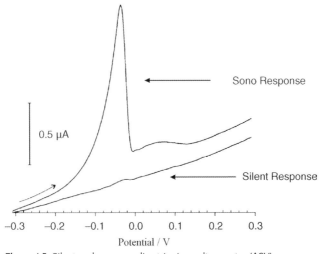

Figure 4.5 Silent and sono anodic stripping voltammetry (ASV) for Cu in industrial effluent recorded at a GC electrode.

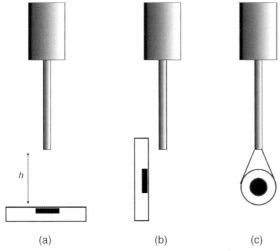

Figure 4.6 Sonic horn alignments: (a) face-on and (b) side-on modes, and (c) a sonotrode.

of water as a reaction environment for organic electrochemical processes, thereby allowing straightforward product isolation by extraction or filtration [57].

4.2
Electrochemical Cell and Experimental Setup

A standard three-electrode electrochemical cell is used for sonoelectrochemical experiments, consisting of working, reference and counter (auxiliary) electrodes. The reference electrode is usually contained in a side arm and is protected from the reaction solution by either a frit or salt bridge, as the reaction medium would typically cause fouling of the reference.

The ultrasound source, taken here to be a probe, can be aligned in two positions relative to the working electrode: face-on or side-on (Figure 4.6).

The most widely used geometry is the face-on mode, where the probe tip is placed opposite the working electrode in the electrochemical cell. The key variables in this arrangement are the electrode–probe separation, h, and the ultrasound power intensity. Another variant of this geometry is where the probe tip is modified to carry an electrode: the so-called "sonotrode," where a single unit acts as both sonic transducer and working electrode (Figure 4.6c).

4.3
Voltammetry Under Insonation

There are several obvious differences between silent and sonovoltammetry (Figure 4.7). The data shown in the figure were obtained for the oxidation of ferrocene at a 2 mm

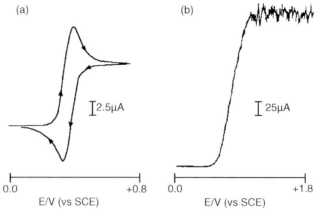

Figure 4.7 Cyclic voltammetry of 2 mM ferrocene in acetonitrile under (a) silent and (b) insonated conditions. Reproduced from [7] with permission from Elsevier.

platinum disk electrode, with an ultrasound power of 50 W cm^{-2} with the electrode in a face-on mode at a distance of 4 mm. The electrochemically reversible ferrocene voltammetry in Figure 4.7(a) is replaced under insonation by a sigmoidal (hydrodynamic) voltammogram with little or no hysteresis (Figure 4.7b) and the mass transport-limited current has increased by around 20-fold. The limiting current is not as stable as with a conventional hydrodynamic electrode, such as a rotating-disk or channel electrode [58], and it contains many irregular spikes whose magnitude is dependent on the power of ultrasound applied.

The appearance of the sonovoltammogram can be understood by consideration of the mass transport conditions experienced by the electrode in the electrochemical cell. In the absence of ultrasound, material arrives at the electrode by diffusion only whereas under insonation both convective flow and cavitation contribute to mass transport. The turbulent convective flow ("acoustic streaming") created by the sonic horn causes average solution speeds of the order of tens of centimetres per second, leading to the sigmoidal appearance of the sonovoltammogram and similarity to those measured for other hydrodynamic systems. In addition, cavitation can occur beyond a certain ultrasound power threshold (dependent on the external applied pressure), and this may preferentially arise at the electrode surface rather than in bulk solution.

4.4
Trace Detection by Stripping Voltammetry

Stripping voltammetry is a highly sensitive method for the electroanalytical detection of a range of metals and some organic substances, achieving very low detection limits that can reach the order of 10^{-10} M in some cases.

4.4.1
Classical Electroanalysis

The methodology is divided into three categories: anodic stripping voltammetry (ASV), cathodic stripping voltammetry (CSV) and adsorptive stripping voltammetry (AdSV). In all cases, there is a pre-concentration step:

$$\text{ASV}: \text{M}^{n+}(\text{aq}) + ne^- \rightarrow \text{M}(\text{electrode})$$
$$\text{CSV}: \text{either } 2\text{M}^{n+}(\text{aq}) + p\text{H}_2\text{O} \rightarrow \text{M}_2\text{O}(\text{electrode}) + 2p\text{H}^+ + 2(p-n)e^-$$
$$\text{or } \text{M} + \text{L}^{p-} \rightarrow \text{ML}^{(n-p)-}(\text{ads}) + ne^-$$
$$\text{AdSV}: \text{A}(\text{aq}) \rightarrow \text{A}(\text{ads})$$

followed by a stripping step where the electrode potential is swept to induce a faradaic current arising from the oxidation/reduction of the accumulated material:

$$\text{ASV}: \text{M}(\text{electrode}) \rightarrow \text{M}^{n+}(\text{aq}) + ne^-$$
$$\text{CSV}: \text{either } \text{M}_2\text{O}(\text{electrode}) + 2p\text{H}^+ + 2(p-n)e^- \rightarrow 2\text{M}^{n+}(\text{aq}) + p\text{H}_2\text{O}$$
$$\text{or } \text{ML}^{(n-p)-}(\text{ads}) + ne^- \rightarrow \text{M} + \text{L}^{p-}$$
$$\text{AdSV}: \text{A}(\text{ads}) + ne^- \rightarrow \text{B}(\text{aq})$$

These methods typically have a high level of accuracy and reproducibility and can be automated by reason of the relative ease of computerization. The required instrumentation is also relatively inexpensive compared to alternative methods such as ICP-MS (inductively coupled plasma-mass spectrometry) or AAS (atomic absorption spectroscopy).

Methods such as ASV or linear sweep voltammetry (LSV) have been applied to the detection of heavy metals such as Bi, Cu, Cd, Mn, Pb, Sn, V and Zn, and also extended to vitamins and pesticides [59–61].

An example of a typical ASV result is for cadmium using a boron-doped diamond (BDD) electrode. The potential at which the peak occurs is specific for cadmium; with other metals ions being stripped at their characteristic potential, this produces a fingerprint for the metal ions in the solution. In this example, the stripping signatures were found to be linear in the range of cadmium additions of 0.2–1.30 μmol dm^{-3} with a limit of detection (LoD) of 2.5×10^{-8} mol dm^{-3}.

However, the technique is limited by the presence in most real-world samples of surface-active impurities that can lead to significant interferences and electrode passivation problems. In many cases, even a molecular monolayer or sub-monolayer may be enough to inhibit the key accumulation step. As a result, more time-consuming pre-treatments are required, rendering the technique less attractive. In addition, ASV has traditionally utilized mercury electrodes, which are increasingly undesirable for environmental reasons. Some of these problems can be addressed by, for example, the use of hydrodynamic techniques, such as the rotating disk electrode to reduce accumulation times, or microelectrodes, but these can suffer from practical problems of having a relatively low limit on rotation speeds and fragility, respectively.

4.4.2
Electroanalysis Facilitated by Ultrasound

The benefits of using ultrasound in electroanalysis have been well documented [23]. Ultrasound was first applied to stripping voltammetry at below the cavitation threshold with the aim of increasing mass transport to extend the sensitivity of ASV [6]. Later, ultrasound was applied beyond the cavitation threshold to overcome electrode passivation in real samples by cavitational cleaning of the electrode surface.

The ultrasound is normally applied during the accumulation step, and switched off prior to the stripping stage where the target analyte is quantified. To illustrate the ability of this method, Figure 4.5 shows a sono versus a silent voltammetric response for stripping analysis of copper in industrial effluent. The only treatment of the sample was to dilute it with 0.1 M perchloric acid and add 100 µL of 1 M potassium chloride. The conventional ASV response under silent conditions gives a small stripping peak that is likely to be highly irreproducible. The response under sonication, however, is a large and easily quantifiable feature. Table 4.1 compares the results from separate effluent samples with other independent analyses, confirming the validity of the sono-ASV method.

4.4.3
Applications of Sono-Anodic Stripping Voltammetry (Sono-ASV)

Early investigations into sono-ASV considered lead and copper contamination of wine and beer, respectively, as examples of media containing large quantities of highly-contaminating organic species [62, 63].

4.5
Biphasic Sonoelectroanalysis

Electroanalysis has been little used in biphasic systems because of difficulties in creating and maintaining an emulsion without surfactants, as these usually would interfere with the analysis. Ultrasound provides a direct means to solve this problem by rapidly establishing and maintaining emulsions of two or more immiscible phases, and has been successfully applied to several systems.

4.5.1
Determination of Lead in Petrol

First, we consider the determination of lead in petrol, which has previously been measured by flame AAS or ICP-MS and DPV (differential pulse voltammetry) [72]. Jagner et al. have used a stripping potentiometric method after pre-treatment [73], and even polarography has been used following an ICl (iodine monochloride) pre-treatment [74]. Lead can be detected more simply using biphasic sono-ASV, and has been reported using a Hg/Pt electrode in an emulsion of the petrol sample with

Table 4.1 Summary of results of sonoelectroanalysis.

Analysis	Method	Ultrasound power (W cm^{-2})	Assay	Detection limit	Independent method	Independent assay	Reference
Pb in wine	Sono-ASV	26	$22 \pm 6 \, \mu g \, L^{-1}$	$2 \, \mu g \, L^{-1}$	AAS	$24 \pm 4 \, \mu g \, L^{-1}$ $27 \pm 4 \, \mu g \, L^{-1}$	[62]
Cu in beer	Sono-ASV	200	$222 \pm 31 \, \mu g \, L^{-1}$ $139 \pm 4 \, \mu g \, L^{-1}$	$32 \, \mu g \, L^{-1}$	AAS	$230 \, \mu g \, L^{-1}$ $140 \, \mu g \, L^{-1}$	[63]
Pb in river sediment	Sono-CSV	14	$187 \, mg \, kg^{-1}$	$10 \, mg \, kg^{-1}$	ICP-MS	$140 \, mg \, kg^{-1}$	[64]
Mn in instant tea	Sono-CSV	14	$1859 \, \mu g \, g^{-1}$ $1859 \, \mu g \, g^{-1}$	$3 \, \mu g \, g^{-1}$	AAS	$1800 \, \mu g \, g^{-1}$ $1000 \, \mu g \, g^{-1}$	[65]
NO_2^- in egg	Sono-LSV	50	$1.2 \pm 0.05 \, mg \, kg^{-1}$ $1.4 \pm 0.05 \, mg \, kg^{-1}$	$0.045 \, mg \, kg^{-1}$	DEFRA data	$1.7 \pm 0.4 \, mg \, kg^{-1}$	[66]
Cu in blood	Sono-ASV	300	$1300 \pm 300 \, \mu g \, L^{-1}$ $620 \pm 60 \, \mu g \, L^{-1}$	$90 \, \mu g \, L^{-1}$	AAS	$1300 \, \mu g \, L^{-1}$ $690 \, \mu g \, L^{-1}$	[67]
Pb in petrol	Sono-ASV	52	$380 \pm 40 \, \mu g \, L^{-1}$	1	AAS	$400 \pm 20 \, \mu g \, L^{-1}$	[68]
Vanillin in food	Sono-biphasic	200	$9.09 \pm 0.21 \, mM$ $9.24 \pm 0.23 \, mM$	$0.020 \, mM$	HPLC	$9.17 \pm 0.21 \, mM$ $9.30 \pm 0.23 \, mM$	[69]
Cu in the presence of 10^3 ppm SDS	Sono-SW/ASV	70	94.1%	$0.3 \, \mu M$	n.d.	n.d.	[70]
Cu in the presence of 10^3 ppm TX-100	Sono-SW/ASV	70	94.1%	$0.3 \, \mu M$	n.d.	n.d.	[70]
Pb in artificial saliva	Sono-ASV	30	Uncontaminated	$0.25 \, \mu g \, L^{-1}$		Uncontaminated	[71]
Pb in human saliva	Sono-ASV	40	$0.92 \pm 0.2 \, \mu g \, L^{-1}$ $5.1 \pm 1.0 \, \mu g \, L^{-1}$	$0.50 \, \mu g \, L^{-1}$	ICP-MS	$0.92 \pm 0.1 \, \mu g \, L^{-1}$ $4.8 \pm 0.5 \, \mu g \, L^{-1}$	[71]
Cd in human saliva	Sono-ASV	40	$2.5 \pm 0.5 \, \mu g \, L^{-1}$ $4.9 \pm 0.05 \, \mu g \, L^{-1}$	$1 \, \mu g \, L^{-1}$	ICP-MS	$2.5 \pm 0.05 \, \mu g \, L^{-1}$ $4.5 \pm 0.05 \, \mu g \, L^{-1}$	[71]

dilute nitric acid [68]. After an initial cleaning step, the accumulation stage was under 50 W cm^{-2} ultrasound for 4 min with the potential held at -1.0 V, after which the Pb was stripped under silent conditions. The following mechanism was then proposed:

Extraction : $\quad$ $PbR_4(org) \leftrightarrow PbR_4(aq)$

Accumulation (-1.0 V) : $\quad$ $PbR_4(aq) + 4e^- + 4H^+ \rightarrow Pb(amalgam) + 4RH$

Stripping : $\quad$ $Pb(amalgam) + 2e^- \rightarrow Pb^{2+}(aq)$

4.5.2
Extraction and Determination of Vanillin [69]

When dissolved in ethyl acetoacetate, vanillin gives oxidative linear sweep voltammetry signals that are dependent on microadditions. In addition the reduction of vanillin in this solvent is nearly reversible, which enables square-wave stripping voltammetry to be performed.

Biphasic analysis has been conducted with ethyl acetoacetate and aqueous vanilla pod extract. The two phases were emulsified and extracted by 1 min insonated at 200 W cm^{-2} prior to square wave voltammetry, although some optimization was required due to electrode passivation effects. Again, independent analysis with HPLC-UV confirmed the accuracy of the ultrasound approach.

4.5.3
Detection of Copper in Blood [75]

The most widely used methods for diagnosing copper deficiency are based on AAS or serum copper content [76]. The former is insensitive to marginal copper deficiency [77] and the latter is expensive and requires sample pre-treatment. There are few other alternatives available [77].

Solvent extraction has been used, but requires time-consuming stirring and separation of the aqueous and organic phases. This can be extended by a double extraction procedure, where the metal ions extracted into the organic phase can then be "reverse-extracted" into a clean aqueous phase before analysis. By using ultrasound for both analytical and extraction steps, the sensitivity and efficiency is improved [75], with N-benzoyl-N-phenylhydroxylamine dissolved in ethyl acetate used as the ligand to extract copper into the organic phase [78]. Sonoemulsification with 1 M acid "reverse-extracts" the Cu into the aqueous phase before analysis is conducted with sono-SWASV. The following scheme was proposed (where HL is the ligand N-benzoyl-N-phenylhydroxylamine):

$HL(org) \leftrightarrow HL(aq)$

$Cu^{2+}(aq) + 2HL(aq) \rightarrow CuL_2(aq) + 2H^+$

$CuL_2(aq) \leftrightarrow CuL_2(org)$

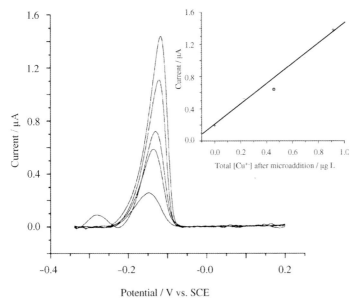

Figure 4.8 Sono-SWASV of Cu from laked horse blood at a GC electrode after insonation. Inset: standard addition plot. With thanks to Dr. C.E. Banks.

It was calculated that a maximum extraction of 74% of copper was possible when HL was in a 1000-fold excess [79].

This method was then successfully applied to laked horse blood [75] and a limit of detection found to be 0.16 µg L^{-1} (Figure 4.8).

4.6
Microelectrodes and Ultrasound

Whilst ultrasound greatly enhances mass transport rates to conventional "macro"-sized electrodes, leading to a steady-state voltammetric response and analytical sensitivity, it may not immediately appear necessary to consider the use of microelectrodes. However, in addition to their even greater rates of mass transport, their size makes it possible to utilize the latest ultrafast instrumentation to probe the fundamental process that occur during insonation. In this section we briefly review some recent results obtained from the application of nanosecond voltammetry to systems under insonation.

4.6.1
Insights into Bubble Dynamics

Microelectrodes have been used to investigate bubble dynamics [21]. Figure 4.9 shows current–time responses for a 29 µm diameter platinum microdisk electrode in

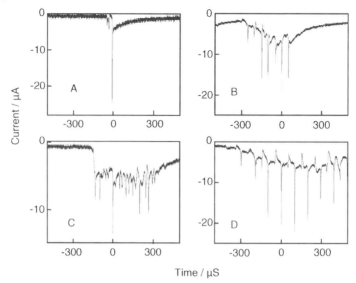

Figure 4.9 Chronoamperograms for the reduction of ferricyanide in aqueous potassium nitrate under an ultrasound power of 9 W cm^{-2}. Reproduced from [21] with permission from the American Chemical Society.

an aqueous solution of potassium ferricyanide under insonation at 20 kHz. The current is monitored at a fixed potential corresponding to transport-limited electrolysis. An ultrafast potentiostat [80] is required for these measurements on the microsecond timescale (and below) so that the response time of the electronics does not distort the current signals.

Close inspection of Figure 4.9 reveals that the spikes are superimposed on a steady background current, and these spikes can be attributed to the formation of bubbles at the electrode surface. The peak of these spikes can be as much as 200 times greater than the steady-state transport-limited current under silent conditions. For disk electrode diameters in the range 25–400 μm the current scales with the electrode area rather than its radius, indicating a switch from convergent diffusion under silent conditions to a near-uniform mass transport regime under insonation. This is due to the effectiveness of acoustic streaming in providing a strong convective flow of material to the electrode. Application of a simple Nernst diffusion model suggests that, for example, a 29 μm diameter microdisk electrode behaves under insonation (10 W cm^{-2}) as a uniformly accessible electrode with a diffusion layer thickness of only 8 μm [21].

Microelectrode arrays have also been used to size bubbles. Maisonhaute et al. [22] have used a five-electrode array of 29 μm independent disks to record bubble-derived signals on the different electrodes (Figure 4.10).

After examining the patterns of current spikes recorded at electrode pairs with differing separations, the authors concluded that bubble sizes could range from as low as 5 to 400 μm and all had very similar magnitude. Accordingly they surmised

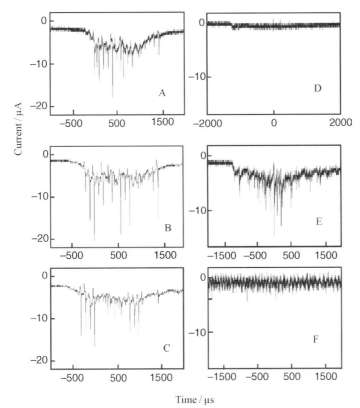

Figure 4.10 Chronoamperograms recorded simultaneously (a–c and d–f) on three electrodes (29 μm diameter) for the reduction of aqueous ferricyanide. Reproduced from [21] with permission from the American Chemical Society.

that the microjetting phenomenon [20] was unlikely to be responsible for the majority of cavitational signals observed, as this would be expected to create a wide variation in spike magnitude depending on whether the electrode was beneath the jet, outside the jet but within the bubble diameter, or outside the cavity [21, 22]. Unfortunately, there was not sufficient time-resolution to identify particular spikes as being common to different electrodes and thereby position and size individual bubbles.

4.6.2
Measurement of Potentials of Zero Charge (PZC) [81]

Further to the investigations into interfacial cavitation, nanosecond chronoamperometry has been applied to the study of biphasic systems, in particular the behavior of an aqueous non-electroactive electrolyte and heptane under variable potentiostatic conditions under acoustic emulsification at 20 kHz. Under sonication, significant current spikes were observed at most potentials. The polarity of the current spikes,

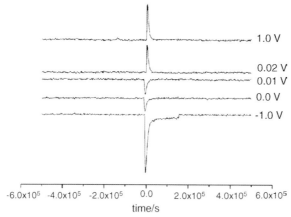

Figure 4.11 Chronoamperograms for heptane droplets in 0.1 M perchloric acid at a Au disk electrode. Reproduced from [81] with permission from the American Chemical Society.

however, was found to invert at a single potential that was specific to the identity of the metal (electrode material) and electrolyte solution. As the inversion potential was approached, the size of the current spikes decreased (Figure 4.11).

The inversion potential coincided with the reported potentials of zero charge (PZC) for both polycrystalline platinum and gold as listed in Table 4.2 (quoted vs NHE for ease of comparison with literature).

The magnitude of the spikes (charge transferred of 10^{-12} to 10^{-10} C) compared to the double layer charge on the 100 μm electrode (5×10^{-9} to 10^{-8} C) suggests that most of the electrode–electrolyte interface is unperturbed by the impact.

4.6.3
Particle Impact Experiments

The direct detection of interactions between single particles and electrode surfaces has concentrated on adhesion events of colloidal particles on mercury electrodes [85–92]. However, the study of time-resolved chronoamperometric signals of single impact events of various solid particles on a polycrystalline platinum electrode in a solution

Table 4.2 Results of oil-droplet impact experiments.

Electrode	Electrolyte	Inversion potential (V vs NHE)	Literature PZC (V vs NHE)
Au	0.1 M HClO$_4$	+0.15 ± 0.01	+0.10 [82]
			+0.14 [83]
Pt	0.1 M KCl/0.01 M HCl	+0.07 ± 0.01	+0.08 [84]
			+0.05 [84]

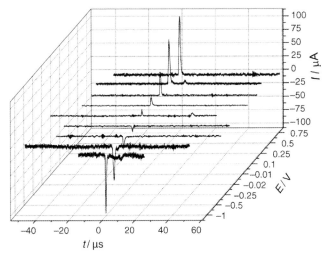

Figure 4.12 Current transients recorded for graphite powder under sonication at different potentials. Reproduced from [93] with permission from the American Chemical Society.

containing only inert electrolyte (i.e., no electroactive species) under conditions of acoustic streaming has been undertaken [93, 94].

The interaction of rigid graphite particles with the electrode surface during acoustic streaming occurs on a drastically shorter timescale than for systems involving nonrigid particles under silent conditions, on the order of 1–5 μs. Under potential control, the polarity of the current spikes was found to invert at a potential corresponding to the potential of zero charge (PZC) for the metal–electrolyte interface and not the impinging graphite particle (Figure 4.12).

The magnitude of the charge transferred during the spikes increases linearly as the potential is moved away from the inversion point (Figure 4.13), until a plateau is reached, representing the maximum charge that can be held by the average particle (in this case, $c.~7 \times 10^{-11}$ C).

It was found by modifying the particles with electroactive species and holding the electrode at a potential sufficient to oxidize/reduce the couple that *no* faradaic charge transfer was observed during a particle impact.

Experiments were also conducted to investigate the effect of different particle materials. Glassy carbon (GC), alumina and calcium carbonate particles were all studied. Figure 4.14 demonstrates that the impact frequency for a given particle size and material is a linear function of the number of particles in solution. However, the results are more complicated for alumina powder and the impact frequencies are about 20 times lower than for the carbon-based materials. The cause of these effects was shown to be the lower conductivity of alumina, since no spikes at all were recorded for particles of the non-conductive $CaCO_3$.

Experiments were then carried out to investigate the relationship between the charge transferred in the current transient and the particle size. Figure 4.15 shows a linear relationship between the transferred charge and the particle radius, which

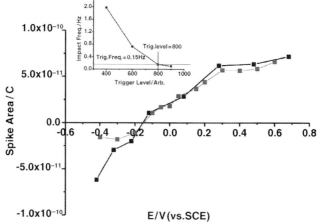

Figure 4.13 Potential dependence of the average spike areas for graphite powder (dark filled squares) and DPPD modified graphite powder (light filled squares). Inset: effect of trigger level on the observed spike frequency. Reproduced from [94] with permission from Wiley-VCH.

indicates that the charge does not depend on the surface area or volume of the sphere (in which cases a squared or cubed dependence would be observed). It was postulated that the charge that the particle can hold depends on the contact region with the (unsmooth) electrode, and that this resembled a spherical cap of surface area:

$$A = 2\pi h r \tag{4.4}$$

where r is the radius of the impacting particle and h is the height of the cap, which would be approximately constant for each combination of electrode and particle materials.

This enabled a probable mechanism to be constructed for the process occurring at the electrode (Figure 4.16). Upon initial contact with the electrode, there is a rapid increase in the magnitude of the current. This is an effect of the large potential

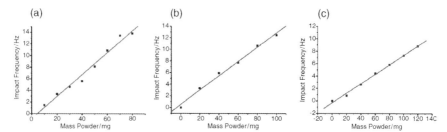

Figure 4.14 Impact frequency as a function of mass of (a) 2–20 μm graphite powder, (b) 20–50 μm spherical carbon powder and (c) 80–200 μm spherical carbon powder. Reproduced from [94] with permission from Wiley-VCH.

4.6 Microelectrodes and Ultrasound | 99

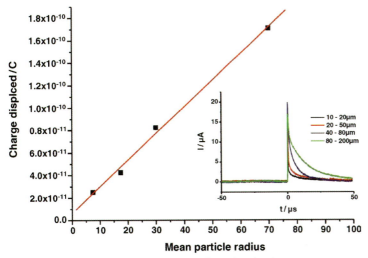

Figure 4.15 Average spike areas as a function of particle radius for glassy carbon spheres. Inset: recorded spikes, which increase with particle size. Reproduced from [94] with permission from Wiley-VCH.

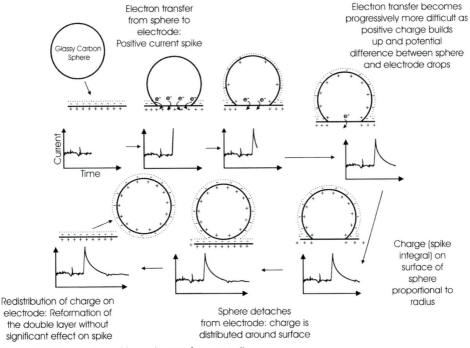

Figure 4.16 Proposed possible mechanism for a sonically induced impact of a spherical particle on an electrode surface. Also shown is the effect on the measured current as a function of time. Reproduced from [94] with permission from Wiley-VCH.

difference between the electrode and the particle. As the charge on the particle increases, the potentials become more similar and the magnitude of the current decreases accordingly. When the current has returned to its original level, the particle has effectively become part of the electrode and the electrical double layer extends over its surface. From the data, it is difficult to say precisely what happens next. However, it is well known that electrodes exposed to ultrasound are much less likely to undergo adsorption process [95]. This principle could be extended to enable us to say that the likely next step in the mechanism would be the detachment of the charged particle into solution. The unidirectionality of the spikes means that there must also be a counter electrode process to maintain charge neutrality.

This work has been extended to the application of particle sizing by using a two-electrode array [96]. Whereas impact events occurring during acoustic agitation have previously been studied in solutions containing only bulk inert electrolyte, this time an electroactive species, ferrocene, was present in the solution (0.1 M tetra-butylammonium perchlorate in acetonitrile) and a suitable potential set to obtain a mass-transport limited oxidation current. If particles are moving within the electro-chemical cell, every particle that contacts with the electrode will perturb the diffusion layer by an amount proportional to the size of the contact area between particle and electrode (and therefore the size of the particle).

Impact events caused two distinct current spikes of different duration. One class of spikes lasted 1–10 μs, as previously reported in non-electroactive solutions, and the other had durations of 30–100 μs, which had a positive polarity (i.e., causing a transient increase in the current above the transport-limited value).

Current spikes were measured at a two-electrode array with a centre–centre distance between the electrodes of 295 μm (measured by optical microscope) for three sizes of spheres: <106 (glass), 188–212 (basalt) and 212–300 μm (glass). The transients recorded were predominantly from single impact events having only been detected at one electrode. However, there were several dual signals, where the same impact was recorded at both electrodes simultaneously (Figure 4.17). Table 4.3 shows the number of impacts recorded in each case.

Of those transients that were only detected at one electrode, the charge transferred during the first 40 μs of the transient was measured to have a maximum value of around 5×10^{-7} C whereas the charge passed for the same electrode geometry in the diffusion-only case of the Shoup–Szabo approximation [97] is approximately 3×10^{-7} C. This indicates that the largest transients correspond to full overlap of particle with electrode and cause complete removal/replenishment of the diffusion layer. The excess charge passed (approx. 67%) is due to compression of the diffusion layer due to turbulent mixing of the particle wake as well as sonication. This is consistent with the incoming particle introducing new electroactive material into the diffusion layer, which is rapidly re-established by convection as well as diffusion.

For the dual impact transients, a simple two-dimensional geometric model can be constructed, assuming that the incoming particle affects the electrode response by its projection onto the electrode plane (i.e., we can treat as a circular "shadow"). Figure 4.18 gives a schematic diagram of the model.

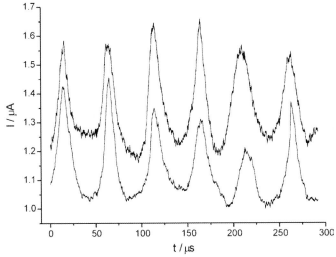

Figure 4.17 Typical trace of multiple impact transients at two electrodes. Note that the periodicity in the impacts is due to the 20 kHz ultrasonic agitation of the solution. The traces have been vertically offset for clarity. Reproduced from [96] with permission from the Royal Society of Chemistry (RSC).

It is then possible to predict a percentage of dual impact events compared to total (or single) impact events as in Table 4.3, based on the size of the impacting particles:

$$\frac{\text{Dual impacts}}{\text{Total impacts}} = \frac{(R+r)\theta - ab}{(R+r)(\pi - \theta) + ab} \quad (4.5)$$

where:

$$\theta = \arcsin\left(\frac{b}{R+r}\right)$$

Therefore, the number of dual impact events can be used as an approximate measure of average particle size. The results in Table 4.3 show good agreement with the commercially determined size distribution.

Table 4.3 Analysis of recorded current transients.

Particle diameter (μm)	Total number of current transients analyzed	Number of pairs of simultaneous current transients	% Dual impacts out of total number of transients	Expected range of % dual out of total number of transients	Effective particle diameter (μm)
<106	201	0	0	0	—
188–212	419	13	3.1	0–1.2	236
212–300	422	30	6.8	1.2–9.8	272

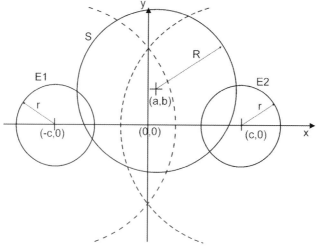

Figure 4.18 Diagram of spherical particle, S, centered on an arbitrary point (a,b), impacting on two electrodes, E1 and E2, sited at (−c,0) and (c,0). A dual impact will only occur if the point (a,b) lies within the region bounded by the dashed lines [the loci of (a,b) for S tangential to E1 and E2]. Reproduced from [96] with permission from the Royal Society of Chemistry (RSC).

4.7
Conclusion

The use of ultrasound in electrochemistry provides a two-way flow of new knowledge and applications. Ultrasound has been used primarily to enhance mass transport and reduce electrode fouling to enable highly sensitive species detection (most notably metal ions), and has some electrosynthetic benefits. However, in recent years, with the arrival of nanosecond voltammetry, it has been voltammetry that has shed new light on ultrasound, with results regarding bubble dynamics and impacts of droplets (in emulsions) and particles (in suspensions), which have provided a relatively simple means of measuring potentials of zero charge and particle sizes.

Abbreviations

Sono-ASV	Sono Anodic Stripping Voltammetry
Sono-CSV	Sono Cathodic Stripping Voltammetry
Sono-LSV	Sono Linear Sweep Voltammetry
Sono-SWASV	Sono Square Wave Anodic Stripping Voltammetry
AAS	Atomic Absorption Spectroscopy
ICP-MS	Inductively Coupled Plasma Mass Spectrometry

DEFRA The UK Government Department for Environment, Food & Rural Affairs
HPLC High Performance Liquid Chromatography

References

1. Suslick, K.S., McNamara, W.B. III and Didenko, Y. (1999) in *Sonochemistry and Sonoluminescence* (eds L.A. Crum, T.J. Mason, J.L. Reisse and K.S. Suslick), Kluwer, Dordrecht, p. 191.
2. Suslick, K.S., Didenko, Y., Fang, M.M. *et al.* (1999) *Philosophical Transactions of the Royal Society of London. Series A: Mathematical and Physical Sciences*, **357**, 355.
3. del Campo, F.J.D., Melville, J., Hardcastle, J.L. and Compton, R.G. (2001) *Journal of Physical Chemistry A*, **105**, 666.
4. Marken, F., Akkermans, R.P. and Compton, R.G. (1996) *Journal of Electroanalytical Chemistry*, **415**, 55.
5. Moriguchi, N. (1934) *Nippon Kagaku Kaishi*, **55**, 751.
6. Marken, F., Rebbitt, T.O., Booth, J. and Compton, R.G. (1997) *Electroanalysis*, **9**, 19.
7. Compton, R.G., Eklund, J.C., Marken, F. *et al.* (1997) *Electrochimica Acta*, **42**, 2919.
8. Hill, H.A.O., Nagakawa, Y., Marken, F. and Compton, R.G. (1996) *The Journal of Physical Chemistry*, **100**, 17395.
9. Birkin, P.R. and Silva-Martinez, S. (1996) *Journal of Electroanalytical Chemistry*, **416**, 127.
10. Leighton, T.J. (1995) *Ultrasonics Sonochemistry*, **2**, S123.
11. Cintas, P. and Luche, J.-L. (1999) *Green Chemistry*, **1**, 115.
12. Gaitan, D.F. and Crum, L.A. (1990) *The Journal of the Acoustical Society of America*, **87**, S141.
13. Berstein, L.S. and Zakin, M.R. (1995) *The Journal of Physical Chemistry*, **99**, 14619.
14. Berstein, L.S., Zakin, M.R., Flint, E.B. and Suslick, K.S. (1996) *The Journal of Physical Chemistry*, **100**, 6612.
15. Lepont-Mullie, F., Pauw, D.D., Lepoint, T. *et al.* (1996) *The Journal of Physical Chemistry*, **100**, 12138.
16. Kwak, H.Y. and Na, J.H. (1997) *Journal of the Physical Society of Japan*, **66**, 3074.
17. Crum, L.A. (1994) *Physics Today*, **47**, 22.
18. Putterman, S. (1995) *Scientific American*, **272** (2), 46.
19. Naude, C.F. and Ellis, A.T. (1961) *Journal of Basic Engineering*, **83**, 648.
20. Crum, L.A. (1979) *The Journal of Physiology*, **11**, C8.
21. Maisonhaute, E., White, P.C. and Compton, R.G. (2001) *The Journal of Physical Chemistry. B*, **105**, 12087.
22. Maisonhaute, E., Brookes, B.A. and Compton, R.G. (2002) *The Journal of Physical Chemistry. B*, **106**, 3166.
23. Banks, C.E. and Compton, R.G. (2003) *ChemPhysChem*, **4**, 169.
24. Klima, J., Bernard, C. and Degrand, C. (1994) *Journal of Electroanalytical Chemistry*, **367**, 297.
25. Klima, J., Bernard, C. and Degrand, C. (1995) *Chemicke Listy*, **89**, 587.
26. Klima, J., Bernard, C. and Degrand, C. (1995) *Journal of Electroanalytical Chemistry*, **399**, 147.
27. Birkin, P.R. and Silva-Martinez, S. (1995) *Chemical Communications*, **17**, 1807.
28. Birkin, P.R. and Silva-Martinez, S. (1997) *Analytical Chemistry*, **69**, 2055.
29. Marken, F., Eklund, J.C. and Compton, R.G. (1995) *Journal of Electroanalytical Chemistry*, **395**, 335.
30. Madigan, N.A. and Coury, L.A. (1997) *Analytical Chemistry*, **69**, 5.
31. Huck, H. (1987) *Berichte der Bunsen-Gesellschaft Physical Chemistry Chemical Physics*, **91**, 648.

32 Jung, C.G., Chapelle, F. and Fontana, A. (1997) *Ultrasonics Sonochemistry*, **4**, 117.
33 Compton, R.G., Marken, F. and Rebbitt, T.O. (1996) *Chemical Communications*, 1017.
34 Hardcastle, J.L., Ball, J.C., Hong, Q. et al. (2000) *Ultrasonics Sonochemistry*, **7**, 7.
35 Montenegro, M.I. (1994) *Research in Chemical Kinetics*, **1**, 1.
36 Amatore, C. (1995) in *Physical Electrochemistry: Principles, Methods and Applications* (ed. I. Rubenstein), Marcel Dekker, New York.
37 Pastore, L., Magno, F. and Amatore, C. (1989) *Journal of Electroanalytical Chemistry*, **26**, 33.
38 Boudjouk, P. (1998) in *Ultrasound: Its Chemical, Physical and Biological Effects* (ed. K.S. Suslick), VCH, Weinheim.
39 Zhang, H. and Coury, L.A. (1993) *Analytical Chemistry*, **65**, 1552.
40 Compton, R.G., Eklund, J.C., Page, S.D. et al. (1994) *The Journal of Physical Chemistry*, **98**, 12410.
41 Peruish, S.A. and Alkire, R.C. (1991) *Journal of the Electrochemical Society*, **138**, 708.
42 Corsaro, R.D., Klunder, J.D. and Jarzynski, J. (1980) *The Journal of the Acoustical Society of America*, **68**, 655.
43 Weissler, A., Cooper, H.W. and Snyder, S. (1948) *The Journal of the Acoustical Society of America*, **20**, 589.
44 Millar, N. (1950) *Transactions of the Faraday Society*, **46**, 546.
45 McLean, R.J. and Mortimer, A.J. (1998) *Ultrasound in Medicine and Biology*, **14**, 59.
46 Mason, T.J., Lorimer, J.P. and Bates, D.M. (1994) *Ultrasonics Sonochemistry*, **1**, S91.
47 Kotronarouu, A., Mills, G. and Hoffman, M.R. (1991) *The Journal of Physical Chemistry*, **95**, 3630.
48 Petrier, C., Jeunet, A., Luche, J.-L. and Reverdy, G. (1992) *Journal of the American Chemical Society*, **114**, 3148.
49 Misik, V. and Reisz, P. (1996) *Ultrasonics Sonochemistry*, **3**, S173.
50 Krishna, C.M., Lion, Y., Kondo, T. and Reisz, P. (1987) *The Journal of Physical Chemistry*, **91**, 5847.
51 Makino, K., Mossoba, M.M. and Reisz, P. (1983) *The Journal of Physical Chemistry*, **87**, 1369.
52 Makino, K., Mossoba, M.M. and Reisz, P. (1982) *Journal of the American Chemical Society*, **104**, 3537.
53 Serpone, N., Terzian, R., Hidaka, H. and Pelizzetti, E. (1994) *The Journal of Physical Chemistry*, **98**, 2634.
54 Gondrexon, N., Renaudin, V., Bernis, A. et al. (1993) *Environmental Technology*, **14**, 587.
55 Hong, Q., Hardcastle, J.L., McKeown, R.A.J. et al. (1999) *New Journal of Chemistry*, **23**, 845.
56 Petrier, C., Jiang, Y. and Lamy, M.-F. (1998) *Environmental Science & Technology*, **32**, 1316.
57 Wadhawan, J.D., Marken, F. and Compton, R.G. (2001) *Pure and Applied Chemistry*, **73**, 1949.
58 Bard, A.J. and Faulkner, L.R. (2001) *Electrochemical Methods: Fundamentals and Applications*, John Wiley, Inc., New York.
59 Banks, C.E. and Compton, R.G. (2003) *Chem Anal (Warsaw)*, **49**, 159.
60 Banks, C.E. and Compton, R.G. (2004) *Analyst*, **129**, 678.
61 Banks, C.E. and Compton, R.G. (2003) *Electroanalysis*, **15**, 329.
62 Akkermans, R.P., Ball, J.C., Rebbitt, T.O. et al. (1998) *Electrochimica Acta*, **43**, 3443.
63 Agra-Gutierrez, C., Hardcastle, J.L., Ball, J.C. and Compton, R.G. (1999) *Analyst*, **124**, 1053.
64 Saterlay, A.J., Agra-Gutierrez, C., Taylor, M.P. et al. (1999) *Electroanalysis*, **11**, 1083.
65 Saterlay, A.J., Foord, J.S. and Compton, R.G. (1999) *Analyst*, **124**, 1791.
66 Davis, J. and Compton, R.G. (2000) *Analytica Chimica Acta*, **404**, 241.
67 Hardcastle, J.L., Murcott, G.G. and Compton, R.G. (2000) *Electroanalysis*, **12**, 559.

68 Blythe, A.N., Akkermans, R.P. and Compton, R.G. (2000) *Electroanalysis*, **12**, 16.
69 Hardcastle, J.L., Paterson, C.J. and Compton, R.G. (2001) *Electroanalysis*, **13**, 899.
70 Hardcastle, J.L., Hignett, G., Melville, J.L. and Compton, R.G. (2002) *Analyst*, **127**, 518.
71 Hardcastle, J.L., West, C.E. and Compton, R.G. (2002) *Analyst*, **127**, 1495.
72 Pournaghi-Azar, M.H. and Ansary-Fard, A.H. (1998) *Talanta*, **46**, 607.
73 Jagner, D., Renman, L. and Wang, Y. (1992) *Analytica Chimica Acta*, **267**, 165.
74 Fontana, A., Braekman-Danheux, C. and Jung, C.G. (1996) *Fuel Processing Technology*, **48**, 107.
75 Hardcastle, J.L. and Compton, R.G. (2001) *Analyst*, **126**, 2025.
76 Subramanian, K.S. and Meranger, J.C. (1983) *The Science of the Total Environment*, **30**, 231.
77 Milne, D.B. (1998) *The American Journal of Clinical Nutrition*, **67**, 1041.
78 Bond, A.M., Nagaosa, Y. and Menjyo, T. (1991) *Analyst*, **116**, 257.
79 Moreno, M.A., Marin, C., Vinagre, F. and Ostapczuk, P. (1999) *The Science of the Total Environment*, **229**, 209.
80 Amatore, C., Maisonhaute, E. and Simonnneau, G. (2000) *Electrochemistry Communications*, **2**, 81.
81 Banks, C.E., Rees, N.V. and Compton, R.G. (2002) *The Journal of Physical Chemistry. B*, **106**, 5810.
82 Bockris, J.O'M., Agrade, S.D. and Gileadi, E. (1969) *Electrochimica Acta*, **14**, 1259.
83 Ataka, K., Yotsuyanagi, T. and Osawa, M. (1996) *The Journal of Physical Chemistry*, **100**, 175.
84 Frumkin, A.N. and Petrii, O.A. (1975) *Electrochimica Acta*, **20**, 347.
85 Charlton, I.D. and Doherty, A.P. (2000) *The Journal of Physical Chemistry. B*, **104**, 8061.
86 Charlton, I.D. and Doherty, A.P. (2000) *Analytical Chemistry*, **72**, 687.
87 Heyrovsky, M., Jirkovsky, J. and Struplova-Bartackova, M. (1995) *Langmuir*, **11**, 4309.
88 Heyrovsky, M., Jirkovsky, J. and Mueller, B.R. (1995) *Langmuir*, **11**, 4293.
89 Heyrovsky, M., Jirkovsky, J. and Struplova-Bartackova, M. (1995) *Langmuir*, **11**, 4300.
90 Hellberg, D., Scholz, F., Schubert, F. *et al.* (2005) *The Journal of Physical Chemistry. B*, **109**, 14715.
91 Scholz, F., Hellberg, D., Harnisch, F. *et al.* (2004) *Electrochemistry Communications*, **6**, 929.
92 Hellberg, D., Scholz, F., Schauer, F. and Weitschies, W. (2002) *Electrochemistry Communications*, **4**, 305.
93 Rees, N.V., Banks, C.E. and Compton, R.G. (2004) *The Journal of Physical Chemistry. B*, **108**, 18391.
94 Clegg, A.D., Rees, N.V., Banks, C.E. and Compton, R.G. (2006) *ChemPhysChem*, **7**, 807.
95 Compton, R.G., Eklund, J.C. and Marken, F. (1997) *Electroanalysis*, **9**, 509.
96 Rees, N.V. and Compton, R.G. (2007) *Analyst*, **132**, 635.
97 Shoup, D. and Szabo, A. (1982) *Journal of Electroanalytical Chemistry*, **140**, 237.

5
Power Ultrasound Meets Protemics

Hugo Miguel Santos, Carlos Lodeiro, and José-Luis Capelo-Martínez

5.1
Introduction

The term proteome usually refers to the entire complement of proteins expressed by a genome, cell, tissue or organism, whilst the term proteomics refers to the study of the proteome (i.e., the complete set of proteins) by using large-scale technologies for protein separation and identification.

The challenge of proteomics is to elucidate the physical organization and to identify and quantitatively monitor the dynamics of proteomes in living organisms as efficiently as possible. With this aim, proteomics is currently applied to the discovery of new protein biomarkers of disease [1], toxicity [2] and drug efficacy [3]. In addition, proteomics is used nowadays for workflow analysis in clinical diagnosis [4].

The past decade has witnessed amazing achievements in proteomics, mainly in the field of instrument development. Nevertheless, the sample treatments used in proteomics remain one of the main limiting steps for rapid protein identification, being in most cases time consuming and with many steps in the sample handling [5]. In fact, it was the introduction of ultrasonic energy that overcame this limitation. As we will see below, sample handling for protein identification used to be tedious, with many single steps, needing as long as 24–48 h in total to complete. However, the application of ultrasonic energy to the different steps of sample treatment for protein identification has allowed common protocols to be carried out in as little as 8 min, with only four main steps, thereby reducing not only the total time necessary but also simplifying the sample handling. This chapter describes, and discusses in detail, the new sample handling protocols for protein identification through peptide mass fingerprint, PMF, and mass spectrometry, MS, based techniques.

Ultrasound in Chemistry: Analytical Applications. Edited by José-Luis Capelo-Martínez
Copyright © 2009 WILEY-VCH Verlag GmbH & Co. KGaA, Weinheim
ISBN: 978-3-527-31934-3

5.2
Protein Identification through Mass-Based Spectrometry Techniques and Peptide Mass Fingerprint

PMF is a methodology that involves the comparison of experimental masses obtained from a pool of peptides originated from a protein (after enzymatic protein digestion) with the peptides produced by the *in silico* (theoretical) enzymatic digestion of the same protein, when present in a database [6]. The experimental masses of the pool of peptides are obtained through mass spectrometry, MS, techniques such as matrix-assisted laser desorption ionization time of flight, MALDI-TOF-MS, or liquid chromatography, LC-MS.

The pool of peptides formed when a protein is digested by an enzyme can be anticipated theoretically due to the selectivity of the enzyme attack on the protein. Some enzymes cleave only certain peptide bonds, producing in this way the same type of peptides for the same protein. Special search programs, known as search engines, are used to compare the experimental masses of the pool of peptides with theoretical ones [7, 8]. The PMF approach has a disadvantage that can not be overcome: the protein is identified only if it is included in a database.

Protein digestion is regularly carried out following two main strategies. In the first approach, named in-gel digestion, a protein is separated from complex mixtures using gel electrophoresis, GE, in the first or second dimension. The piece of gel, called a spot, that contains the protein (sometimes the proteins) is then excised and a complicated sample treatment is performed to degrade the protein(s) contained in the spot. The pool of peptides so-obtained is then used for PMF. In the second strategy, named in-solution digestion, the proteins are separated by LC. The protein(s) are then digested in-solution, and the pool of peptides, as in the in-gel digestion process, is used for PMF. Several permutations of these two approaches have been reported from time to time.

The GE method is well known for its resolving power, massively parallel separation, quantitative nature and instant visualization of thousands of protein species, including post-translationally modified protein isoforms [9]. A lack of reproducibility and its labor-intensive nature are the main claimed drawbacks. The LC approach is based on variations of multidimensional chromatography, that is, on on-line approaches using more than one chromatographic column [10]. This approach takes a long time due to the nature of the separation process. Regardless of the differences between the two methods, the selection of either one depends on user preferences and they should be considered complementary.

5.3
Classic In-Gel Protein Sample Treatment for Protein Identification through Peptide Mass Fingerprint

Figure 5.1 shows a comprehensive scheme for the GE of proteins in the first dimension by the classic overnight sample treatment and by the same treatment

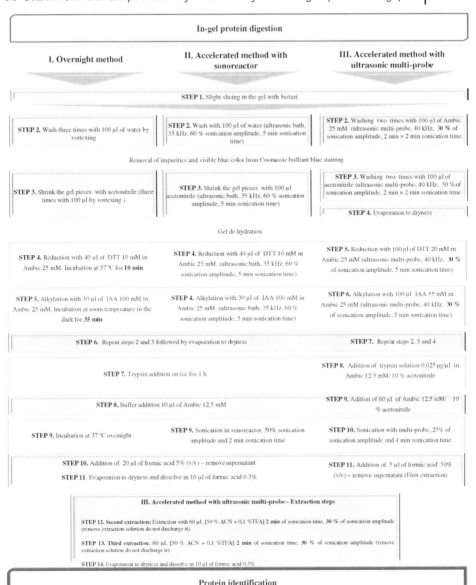

Figure 5.1 Comprehensive scheme for in-gel protein sample treatment for protein identification by spectrometry based techniques and peptide mass fingerprint. I: overnight method (16 h); II: accelerated method with sonoreactor (2 h); and (III) accelerated method with ultrasonic multi-probe (1 h).

accelerated with the aid of ultrasonic energy. As can be seen, many steps must be followed to digest a protein correctly. When the proteins are separated through two-dimensional GE the number of steps increases, and the method becomes more tedious and more time-consuming [11] than the one shown in Figure 5.1.

After separation and purification the protein complex mixture is usually denatured in a mixture of 8–9 M urea/thiourea prior to 1D-GE separation. Denaturation of the protein is mandatory to obtain an efficient, reproducible separation in the gel. In addition, thiourea is added to increase the solubility of hydrophobic proteins [12].

When the first dimension is performed in a two-dimensional scheme, the sample is submitted to isoelectric focusing (IEF) gel strips containing ampholytes, which are substances that create a pH gradient along the gel. Reproducibility of the protein separation in modern gel strips is obtained through the immobilization of ampholytes in the form of acrylamide derivatives, called immobilines [13].

Once the first dimension has been completed, the separated proteins must be protected against further re-oxidation. To do so, protein cystine residues must be reduced using dithiothreitol (DTT) and the resulting cysteines blocked with iodoacetamide (IAA).

For a 2D-GE separation, the strips are soaked in SDS (sodium dodecyl sulfate) to prepare proteins for separation by molecular weight (MW) in the second dimension.

After separation by the one- or two-dimensional approaches, the proteins must be visualized in the gel. Different staining protocols can be found in the literature [14, 15], including the most popular ones, that is, Coomassie blue and silver nitrate methods. Modern stains for protein visualization, such as sypro red or sypro orange fluorescent dyes, are less intensive-labor in terms of handling than Coomassie blue or silver nitrate [15].

Once visualized in the spots, the proteins are treated in two different forms. They can be eluted in intact forms from the gel by passive elution, electroelution or using ultrasonic energy [16–18]. After elution, proteins are in-solution digested and subsequently identified by PMF. Protein elution from gels, nevertheless, is not easy and, indeed, is not mandatory for protein identification by PMF – it only adds more handling to the sample treatment.

Proteins can be digested *in situ*, that is within the gel. The advantage of this procedure over protein elution is that peptides formed during the digestion process are eluted more easily from the gel than are the whole proteins.

Focusing on the in-gel digestion process, after protein visualization, the spot is excised and submitted to a de-staining process to remove the visualization agent. This is done because the staining chemicals may affect the subsequent steps of the PMF process, especially the MS determination. After de-staining the gel, the enzyme must be added to begin protein digestion. An adequate enzyme/substrate ratio is needed to avoid low digestion efficiency. It is also important to guarantee that the enzyme is delivered inside the gel, where the protein actually is. To do so, gel slides are first dehydrated using acetonitrile and dried in a speed-vacuum. A solution containing the enzyme is then added to the dehydrated gel. During the subsequent re-hydration process the enzyme is transported inside the gel by the solution. The solution should be buffered to maintain the pH of the digestion process to its optimum value to ensure maximum trypsin activity. In addition, the temperature of the bulk solution during enzymatic digestion needs to be controlled for correct enzymatic performance. Incubation for a few hours to overnight at 30 or 37 °C is usually sufficient for complete protein digestion.

Once protein digestion is completed, the supernatant containing the peptides is acidified to stop enzymatic activity, to prevent the enzyme digesting itself, which would give an excess of interfering peaks in the subsequent MS measurement.

The MS analysis provides an experimental number of m/z peaks corresponding to peptides, which are compared with the theoretical ones assigned by the search engine as the most probable ones matching the experimental data. The relation between both sequences, expressed as a percentage, is known as the sequence coverage of the PMF identification process.

The in-gel digestion process has some drawbacks. First, since protein is trapped in the gel the enzyme can not freely access all the protein structure – thus some peptide bonds remain inaccessible to the enzyme. Second, not all peptides produced during digestion can diffuse freely from the gel.

As the comprehensive scheme depicted in Figure 5.1 shows, many different steps, including washing with water and acetonitrile, need to be carried out to complete the sample treatment. Overall, from 4 h at 37 °C to 24 h is required to complete only the enzymatic digestion step. As we will see below, the introduction in 2005 of ultrasonic energy as a way to accelerate de different sample treatment steps (shown in Figure 5.1) has allowed the whole process to be completed in only 1 h [26]. In addition, the labor-intense handling has been enormously simplified.

This section has described a general procedure for the in-gel protein digestion process, variants of which can be found in the literature.

5.4
Ultrasonic Energy for the Acceleration of In-Gel Protein Sample Treatment for Protein Identification through Peptide Mass Fingerprint

5.4.1
Washing, Reduction and Alkylation Steps

As depicted in Figure 5.1, prior to enzymatic digestion the gel band containing the protein must be washed with water to remove contaminants and with acetonitrile to dehydrate the gel and to remove staining agents. Then, in-gel proteins are submitted to reduction and alkylation steps to facilitate the subsequent trypsin action. Finally, another washing step is required. The total time involved in these steps is about 3 h. Although some of these steps can be carried out by commercial robots, thus simplifying the handling performed by the operator, the high costs of such apparatus is an important limitation for many laboratories.

For simplicity, the treatment depicted in Figure 5.1 can be considered as consisting of four main steps: (i) washing, (ii) reduction, (iii) alkylation and (iv) digestion.

Concerning the washing procedure, common protocols recommend cleaning the gel by washing it three times with water plus three times with acetonitrile. Taking into account that the washing procedure must be done before and after the reduction and alkylation steps, the total time required is about 90 min. When the washing procedure is accelerated and improved with ultrasonic energy, it is achieved in only 20 min in a single step with water, followed by another single step with acetonitrile [19]; without protein degradation caused by ultrasound. In addition, the number of peptides matched and the protein sequence coverage reported were similar or even better than the ones obtained using the classic protocol without ultrasonication for the following proteins: glycogen phosphorylase b, BSA, ovalbumin, carbonic anhydrase, trypsin inhibitor and α-lactalbumin. Furthermore, specific proteins have been identified from complex protein mixtures obtained from three different sulfate-reducing bacteria: *Desulfuvibrio desulfuricans* G20, *Desulfuvibrio gigas* NCIB 9332 and *Desulfuvibrio desulfuricans* ATCC 27774. The acceleration study of the washing procedure was carried out with three different ultrasonic energy sources: the ultrasonic bath, UB, the ultrasonic probe, UP and the sonoreactor, SR. Interestingly, the UP and SR can reduce the time required to 2 min while the UB needs 5 min. However, the throughput obtained with the bath cannot be reached with the sonoreactor or with the probe. Finally, it must be stressed that not only time was saved but, also, the sample handling was simplified using ultrasonication.

Reduction and alkylation steps, as shown in Figure 5.1, are normally carried out to avoid protein renaturation, making the enzymatic digestion easier. During those steps, cystine residues are reduced using dithiothreitol (DTT) for 10 min at 60 °C plus 15 min at room temperature, and the resulting cysteines are blocked with iodoacetamide (IAA) for 35 min in the dark at room temperature. However, a study of the possible accelerating effects of the three ultrasonic devices mentioned above concluded that the time taken for the reduction and alkylation steps can be lowered from 60 to 10 min using the bath and to 4 min using the sonoreactor or probe [19]. The number of peptides matched and the protein sequence coverage were not affected by the sonication process.

An interesting finding is related to the step in which trypsin is added on ice to the gel and left to stand for 60 min. The ice bath is needed to lower the temperature of the solution so that trypsin autolysis is avoided. Although this step could not be accelerated using the SR [19], the use of an UP allowed this step to be avoided, making possible the rapid digestion of the proteins [20]. Rather than a lack in the efficiency of the ultrasonic energy, the failure of the SR in accelerating the process seems to be related to the solution in which the trypsin was dissolved. Thus, while the trial with the sonoreactor was done with trypsin dissolved in ammonium bicarbonate buffer, AMBIC, 12 mM, with the UP was performed with trypsin dissolved in a mixture of water/acetonitrile (ACN).

Overall, the previously time-consuming and tedious sample handling enzymatic digestion, for identification of proteins separated by GE using PMF, can be simplified by the introduction of ultrasonic energy in the different steps of the sample treatment. The total time is reduced by *c.* 85% without compromising the protein sequence coverage or the number of peptides matched. In addition, the sample handling is drastically simplified. Furthermore, no increase in the background of the MALDI spectra was observed. The acceleration of the different stages, washing, protein reduction and protein alkylation, in the sample treatment for protein identification by PMF can be carried out with similar results using any of the following ultrasonic devices: bath, probe or sonoreactor. However, higher sample throughput is obtained using the bath, and so it is the recommended device for speeding up the washing, reduction and alkylation steps.

5.4.2
In-Gel Protein Digestion Process

Once the gel piece has been submitted to the steps of reduction and alkylation, it is necessary to perform the protein digestion. This step used to be carried out overnight, taking as long as 12 h or with samples that are difficult to digest, such as membrane proteins, it can even take 24 h [6]. As we will see below, different attempts have been made to speed up this step; they are based on increasing the temperature during the digestion step [21], or using columns containing immobilized trypsin [22], combinations of these [23], addition of organic solvents [24] or enhancing trypsin digestion by microwave energy [25]. Notably, ultrasonic energy can, theoretically, be incorporated in any of the these attempts.

Since the introduction in 2005 [26] of ultrasonic energy to speed up the in-gel enzymatic digestion step of proteins for protein identification by PMF, many improvements and findings have been made [15, 19, 27–31]. Although the mechanism responsible for the enzymatic digestion enhancement using focused ultrasound is not completely understood yet, it appears to be related to an increase in diffusion rates as a consequence of cavitation phenomena. Research in ultrasonic applications in medicine and drug delivery has estimated that the pressure at the tip of the jet generated by bubble collapse (cavitation phenomena) is around 60 MPa [32]. Hence, liquid jets may act as microsyringes, delivering the enzyme to a region of interest, inside the gel, thus facilitating digestion of proteins. Furthermore, when a cavitating bubble collapses near the surface of a solid particle, micro-jets of solvent are propagated toward the surface at velocities greater than $100\,\mathrm{m\,s^{-1}}$, causing mechanical erosion of the solid surface, leading to solid disruption [33]. This process could be favorably used to enhance peptide release from the gel.

The following subsections deal with the variables that affect the performance of this step.

5.4.2.1 Sample Handling
It must be emphasized that all ultrasonic devices can neither equally speed up the activity of the enzymes nor enhance in-gel protein digestion. This difference can be related to the ultrasonic intensity delivered for each type of ultrasonic system. Thus, for a constant volume of 1.5 mL, the intensity of sonication delivered is 0.01 W for the ultrasonic bath, 0.5 W for the sonotrode and 15 W for the ultrasonic probe [34]. The ultrasonic bath cannot accelerate the digestion process to times as short as obtained with the sonotrode or the sonoreactor, and should not be used for this purpose.

In contrast, the sonoreactor can be used to handle up to six samples at once, and is the recommended apparatus to perform the protein digestion step [29]. The ultrasonic probe also gives good results [26–28]. Handling, however, is more complicated than with the sonoreactor. Thus, the size of the probe that can be used is limited by the sample volume to be treated, and must be chosen carefully. For example, a 0.5 mm probe is used to deliver ultrasonic energy in a volume of 10–500 µL whilst a 1 mm probe can deliver ultrasound in a volume range of 100–5000 µL [26, 27]. The paramount importance of maintaining the gel slide under the ultrasonic field generated by the probe has also been reported. The existence of the so-called "dead zone" (i.e., the zone where cavitation is not achieved) is one of the most important factors to consider when using probe sonication (Figure 5.2). Dead zones are related to the distribution of ultrasonic energy, since variations in the local cavitational activity and the resulting pressure field as a function of axial and radial distance from the probe have been demonstrated [35, 36]. Hence, the distance between the probe and the wall container must be as short as possible. Eppendorf-type vessels should be employed, since the small diameter raises the liquid level of the sample without increasing the volume, thereby allowing the probe to be inserted deep enough into the solution. As shown in Figure 5.2, with the 0.5 mm probe the gel pieces tend to reach the "dead zones" of the ultrasonic process, where the cavitation effects are negligible. Under these conditions, enzymatic digestion does not take place. This problem is not observed for the 1 mm probe, since the excised gel does not have enough free space to reach the dead zone, remaining fully exposed to the ultrasonic field during treatment (Figure 5.2); under these conditions, protein digestion was successfully attained [26].

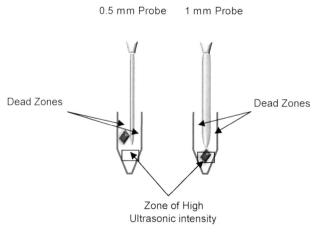

Figure 5.2 Scheme showing the correct application of an ultrasonic probe for in-gel protein digestion. For the 0.5 mm probe, the gel is out of the area of high intensity sonication, and it has been displaced to a dead zone. For the 1 mm probe the gel is positioned correctly.

5.4.2.2 Sonication Volume

The sonication process can be performed immediately after the gel spot has been completely covered by the digestion solution. In this sense the sonoreactor permits treatment with a volume as low as 25 µL. Although the UP can also be used with this low volume, it must be considered that a liquid sonicated with this device always produces an aerosol. Consequently, small drops of the solution are spread through the walls of the container. When a small amount of liquid is sonicated, such as 25 µL, such aerosol formation becomes problematic since a considerable fraction of the liquid is deposited on the container's wall, away from the ultrasonic effects. This forces the user to stop the sonication process each 20 s and use centrifugation to force the solution back down to completely cover the gel. To avoid this labor-intensive handling, the minimum volume recommended when working with UP is 100 µL.

5.4.2.3 Sonication Time

As a general rule, low sonication times are needed to perform the in-gel enzymatic digestion process. The more intense the sonication device the shorter the total time required. The UP allows completion in less than 60 s [26]. However, longer times (120 s) can also be used [28]. With the sonoreactor, longer times are needed to complete the digestion processes. These times are, however, below 5 min [29]. When this approach is carried out with an unknown sample, it must be borne in mind that excessive sonication time can degrade the proteins, promoting undesired cleavages, leading to poor protein coverage and, ultimately, to failure in the protein identification process [28].

5.4.2.4 Sonication Amplitude

As explained in Chapter 1, cavitation effects are linked directly to the sonication frequency and amplitude. For the same frequency, higher amplitudes lead to intensification, to some extent, of the cavitation effects. Studies varying the sonication amplitude between 20 and 80% have been reported for BSA protein and for α-lactalbumin [28]. Whilst for the latter protein worse results were obtained for high frequencies, no significant variations were found for BSA, in the amplitude ranges studied, neither in protein coverage nor in the number of peptide identified. Hence for some proteins, especially ones with low masses such as α-lactalbumin, high amplitudes must be avoided. As a general rule, a maximum of 50% sonication amplitude is recommended for both the SR and the UP.

5.4.2.5 Protein to Trypsin Ratio

Since the enzyme is responsible for protein digestion, the optimization of its amount is of paramount importance. Relevant data reported on this matter reveal the interesting conclusion that the use of ultrasonication does not allow a reduction in the amount of trypsin employed, no matter what type of ultrasonic device is used [26–29]. This suggests that ultrasonic energy enhances the kinetics of the enzymatic digestion of proteins rather than increases the trypsin efficiency in terms of cleavage capability. In other words, the application of ultrasonic energy does not make trypsin be able to digest more protein, the digestion process itself is just accelerated. For this reason the recommended protein to trypsin ratio is 20 : 1 w/w, the same as that recommended by most manufacturers. It must be considered that the protein cleavage catalyzed by trypsin is thermodynamically favorable but requires significant activation energy, which is kinetically unfavorable. Ultrasonication helps to speed up kinetic reactions by promoting more enzyme to substrate contacts per second, and increasing diffusion coefficients. This can explain the favorable digestion under the effects of an ultrasonic field, and it also helps in understanding why ultrasonic baths can not accelerate the enzymatic kinetics to times as short as those obtained with the SR and UP. The UB is so inefficient in terms of ultrasonic energy delivery that the activation energy is not increased sufficiently to speed up the enzymatic kinetics.

5.4.2.6 Temperature

It seems that temperature control of the protein digestion is not always of great concern for sonication times up to 120 s [27]. Since the time generally recommended for the SR is 4 min, it is necessary to change the water used every 2 min to avoid excess heating of the sample. It must be remembered that high temperatures can inactivate enzymes by denaturation, thus stopping the digestion process.

5.4.2.7 Solvent

Different solutions have been used to test the influence of the composition of the solution on the efficiency of the digestion process under the effects of an ultrasonic field. Results are worst (lower number of peptides identified) when the digestion is performed in water, without buffering the media. Maintaining an adequate pH to obtain the correct enzyme activity seems to be important [26] and it is recommended. The utilization of solvent mixtures of water/organics has been reported to help in obtaining better results; however, the use of ultrasonication and a mixture of water/methanol (80 : 20 v/v) did not improve the digestion [26]. Nevertheless, the utilization of a water/ACN (50 : 50 v/v) solution gave equivalent results, in terms of number of peptides identified and sequence coverage, to the classic protocol or the accelerated protocol with ultrasound and buffered solutions [37]. Another advantage of using a water/ACN solution is that the signal-to-noise ratio in the MALDI is better than in approaches using buffered solutions. This is due to the absence of salts. As shown below, this finding has allowed simplification of the sample handling used in the in-solution digestion protocol [38].

5.4.2.8 Minimum Amount of Protein Identified

The minimum amount of protein that can be identified using the ultrasonic protocol or the non-ultrasonic one is the same, for example 0.1 µg for a trypsin concentration of 0.6 µM [28]. In these results the ultrasonic device used had no influence, with the lower amount of protein detectable being equal for the SR and UP. As explained above, ultrasonication is not able to increment the amount of protein digested for a constant amount of trypsin – it simply assists the digestion process by making it more rapid.

5.4.2.9 Reduction and Alkylation Steps

Several reported trials have attempted to eliminate the protein reduction and protein alkylation steps by performing protein digestion with the aid of ultrasonic energy only. Unfortunately, the results were unexpectedly bad. In fact the sequence coverage was half that obtained for the same conditions of treatment but with the reduction and alkylation applied in the sample treatment [27]. Therefore, it is recommended that the reduction and alkylation steps always be performed.

5.4.2.10 Comparison with Other Types of Rapid Sample Treatments

Some authors have suggested two additional ways to speed up enzymatic digestions: heating and microwave energy. It has been claimed that heating the solution and gel spot at 60 °C for 30 min is enough to obtain similar results, in terms of protein digestion, to the overnight procedure [21]. Heating has been compared with ultrasonic energy [26]; for both methods comparable signal-to-noise ratios and peptide intensities, number of peptides identified and protein coverage were observed. Concerning microwave energy, some preliminary results carried out in our laboratory suggest a better performance for the ultrasonic protocol in the digestion process.

5.4.2.11 Influence of Protein Staining

In GE, selection of the appropriate protein staining to visualize the proteins separated in the gel is of great concern, because it is linked directly to the quality of the results obtained. The type of protein staining used is decisive in choosing (i) the subsequent sample treatment in the GE (e.g., there are at present over 100 different variants of silver-staining protocols) and (ii) the technique used to detect the mass of the peptides, namely, MALDI-TOF-MS or HPLC-MS/MS.

Notably, in ultrasonic treatments for in-gel protein identification by PMF, the digestion process is accelerated by using a SR or an UP. One effect of ultrasonic energy is to extract other substances present in the gel, during the digestion process. Such substances can be a serious problem for the mass spectrometry measurements. For this reason, the compatibility of ultrasonic treatments and chemicals for protein staining has been studied [15]. The following staining protocols were studied under the effects of an ultrasonic field: Coomassie blue and silver nitrate, both visible stains, and the fluorescent dyes sypro red and sypro orange. The SR was the ultrasonic device chosen to perform these studies because it has been observed that the UP, when used for times longer than 4 min, can lead to degradation of the gel spot in such a way that the solution obtained can block the chromatographic column used to separate the peptides in HPLC-MS/MS. Similar results for the identification of six different proteins were obtained using the overnight protocol or the ultrasonic one with any of the four staining methods studied. For the Coomassie blue method, a sonication time in the digestion step of 2 min was enough, while the silver nitrate, sypro red and sypro orange methods required 4 min sonication to attain the same protein coverage and number of peptides identified by the overnight method.

Protein reduction and protein alkylation steps must be included in the overnight and in the ultrasonic protocols regardless of the staining method studied, to achieve good results, in terms of protein sequence coverage and number of peptides matched. Moreover, the sensitivity of the method can be improved by increasing the number of peptides extracted – by carrying out several extractions from the gel with a mixture of acetonitrile/TFA/water, once the enzymatic digestion has been completed.

Table 5.1 shows a comparison of classic and accelerated in-gel protein methods for six standard proteins and three proteins from a complex mixture. As can be seen, similar results were obtained for both methodologies in terms of peptides identified and sequence coverage.

5.5
Classic In-Solution Protein Sample Treatment for Protein Identification through Peptide Mass Fingerprint

Shotgun proteomics, first developed in 1999 [10], is based on the in-solution digestion and analysis of proteins contained in a whole-extract. This procedure has been used extensively in the analysis of isolated proteins, but nowadays is applied to complex mixtures of proteins.

Protein composition and characteristics are very diverse. As consequence, the performance of the protein digestion, the number, type and quantity of peptides obtained differ greatly from one protein to another. Generally, more abundant proteins are usually easier to identify. Nevertheless, protein biomarkers, that is, proteins with potentially clinical relevance, belong to the less abundant proteins and consequently, when the target protein is in this group, protein depletion is a mandatory step. Protein depletion is performed to avoid interferences caused by the most abundant proteins in the detection of less abundant ones [39]. Finally, two other problems associated with protein identification in complex mixtures are (i) from sample to sample the matrix can be heterogeneous, making results irreproducible and difficult to interpret, and (ii) the quantity and number of proteins can differ from sample to sample. To overcome these problems, when dealing with complex mixtures, different approaches can be found in literature, but essentially all respond to the same scheme (Figure 5.3). First, a more or less selective protein depletion step is done. The precipitate or the supernatant can be used further for investigation. In any case, proteins are then dissolved (or re-suspended) in a buffered media, generally ammonium bicarbonate, supplemented with urea or any other chaotropic agent. This media helps to break intramolecular forces and helps to denature proteins. As explained above, protein denaturation is of primary importance to help in the subsequent enzymatic digestion process. In fact, to prevent protein renaturation before enzymatic attack, reduction and alkylation of protein disulfide bridges is commonly carried using DTT and IAA, respectively, as in the in-gel protein digestion protocol discussed in previous sections.

5.5 Classic In-Solution Protein Sample Treatment for Protein Identification

Table 5.1 Sequence coverage and number of unique identified peptides for the in-gel classic protocol and in-gel ultrasonic (US) treatment[a].

Protein	Database	Classic protocol		US bath/SR	
		Sequence coverage (%)	No. of unique peptides identified	Sequence coverage (%)	Number of unique peptides identified
Glycogen phosphorylase	SwissProt	65.5 ± 0.7	64.0 ± 1.4	64.5 ± 2.1	64.0 ± 1.4
BSA	SwissProt	73.0 ± 1.4	50.5 ± 2.1	65.5 ± 3.5	42.0 ± 1.4
Ovalbumin	SwissProt	49.0 ± 5.7	20.0 ± 1.4	53.0 ± 1.4	22.0 ± 2.1
Carbonic anhydrase	SwissProt	69 ± 0	20 ± 0	64 ± 0	19.5 ± 0.7
Trypsin inhibitor	SwissProt	51.0 ± 1.4	18 ± 0	53.0 ± 1.4	18.5 ± 0.7
α-Lactalbumin	SwissProt	51 ± 0	11 ± 0	45.4 ± 7.8	9.5 ± 0.7
Sulfite reductase (EC 1.8.99.1), *Desulfovibrio desulfuricans*	MSDB	40.0 ± 1.4	17.5 ± 3.5	42.5 ± 3.5	20.5 ± 0.7
Zinc resistance-associated protein precursor, *Desulfovibrio desulfuricans* (strain G20)	MSDB	37.5 ± 6.4	13.5 ± 0.7	46.5 ± 2.1	13.0 ± 0
Sulfite reductase, dissimilatory-type subunit alpha (EC 1.8.99.3) (Desulfoviridin subunit alpha) (hydrogen sulfite reductase alpha subunit) (fragment)	SwissProt	47.8 ± 2.5	15.5 ± 0.7	56.1 ± 4.3	21 ± 2.8

[a] Ultrasonic method: cleaning, reduction and alkylation were carried out in 5 min sonication time using the ultrasonic bath (100% sonication amplitude) while the digestion was performed with the sonoreactor (2 min sonication time and 50% sonication amplitude) ($n = 2$). Protein amounts ranged from 0.1 to 1 μg.

In-solution protein digestion

Overnight method	Accelerated method with urea	Accelerated clean method with acetonitrile

STEP 1. 20 µl of protein solution in urea 6.5 M		STEP 1. 20 µl of protein solution in H_2O/ACN

| Denature agent [a] – urea
[a] Denaturation of proteins involves the disruption and possible destruction of both the secondary and tertiary structures. Denaturation occurs because the bonding interactions responsible for the secondary structure (hydrogen bonds to amides) and tertiary structure (bonding interactions between "side chains" including: hydrogen bonding, salt bridges, disulfide bonds, and non-polar hydrophobic interactions) are disrupted. | | STEP 2. 1 min of sonication time with sonoreactor and 50% of amplitude |
| | | Denature agent – Acetonitrile, denaturation effect is enhanced by ultrasonication [a] |

| STEP 2. Reduction with 2 µl of DTT 110 mM in Ambic 12.5 mM. Incubation at 37 °C for **60 min** | STEP 2. Reduction with 2 µl of DTT 110 mM in Ambic 12.5 mM. **5 min** of sonication time with sonoreactor and 50% of sonication amplitude | STEP 3. Reduction with 2 µl of DTT 110 mM in Ambic 12.5 mM. **1 min** of sonication time with sonoreactor and 50% of sonication amplitude |

Reduce the dissulfide bonds of proteins and, more generally, to prevent intramolecular and intermolecular disulfide bonds from forming between cysteine residues of proteins

| STEP 3. Alkylation with 2 µl of IAA 600 mM in Ambic 12.5 mM. Incubation at room temperature in the dark for **45 min** | STEP 3. Alkylation with 2 µl of IAA 600 mM in Ambic 12.5 mM, **5 min** of sonication time with sonoreactor and 50% of sonication amplitude | STEP 4. Alkylation with 2 µl of IAA 600 mM in Ambic 12.5 mM, **1 min** of sonication time with sonoreactor and 50% of sonication amplitude |

Alkylation of the SH groups, the cysteines are transformed to the stable S-carboxyamidomethylcysteine (CAM; adduct: -CH_2-$CONH_2$). This chemical modification allows for proteins with a high number of disulfide bonds the successful identification as well as the highest peptide yield and sequence coverage.

STEP 4. Buffer addition (72 µl of Ambic 12.5 mM)		STEP 5. Buffer addition (72 µl of Ambic 12.5 mM)
STEP 5. An aliquot of the sample (10 µl) was transferred to another eppendorf		STEP 6. An aliquot of the sample (10 µl) was transferred to another eppendorf
STEP 6. Trypsin addition (protein enzyme ratio 20:1 (w/w)). Incubation at 37 °C overrnight	STEP 6. Trypsin addition (protein enzyme ratio 20:1 (w/w)). 5 min of sonication time with sonoreactor and 50 % of sonication amplitude	STEP 7. Trypsin addition (protein enzyme ratio 20:1 (w/w)). 5 min of sonication time with sonoreactor and 50 % of sonication amplitude

Trypsin predominantly cleaves peptide chains at the carboxyl side (or C-terminal side) of the amino acids lysine and arginine, except when either is followed by proline. The aspartate residue (Asp 189) located in the catalytic center of trypsin is responsible for attracting and stabilizing positively-charged lysine and/or arginine, and is thus responsible for the specificity of the enzyme. Trypsin have an optimal operating pH of about 8 and optimal operating temperature of about 37°C.

STEP 7. Addition of 1µl of formic acid 50% (v/v)		STEP 8. Addition of 1µl of formic acid 50% (v/v)
STEP 8. Zip Tip desalting procedure		STEP 9. No desalting is needed

Protein identification

Figure 5.3 Comprehensive scheme for in-solution protein sample treatment for protein identification by spectrometry based techniques and peptide mass fingerprint. Overnight method (16 h); accelerated method with ultrasonication and urea (25 min); and clean method accelerated with ultrasound (8 min).

Protein digestion, as shown in Figure 5.3, is generally done using trypsin, but some in-solution protocols, however, make use of two digestion steps. In the first step the enzyme lysine is used, because it can work in the presence of 8 M urea (the common urea concentration used to denature proteins). In the second step the protein is further digested using trypsin [40].

The digestion process is stopped by lowering the pH of the solution, thus inactivating the enzyme. Further sample treatment can accomplish lyophilization or evaporation in a speed vacuum to increase sample concentration or for sample storage. A treated sample needs to be stored at below $-20\,°C$ or, if possible, at below $-70\,°C$.

Protein identification is the final item of the in-solution sample digestion protocol. It can be achieved by PMF using MALDI-TOF-MS or using HPLC coupled to electrospray ionization (ESI)-MS/MS. With MS/MS, proteins are identified through peptides that are further fragmented in the MS/MS spectrometer. This means of protein identification is known as the peptide-fragment fingerprint (PFF) method.

5.6
Ultrasonic Energy for the Acceleration of the In-Solution Protein Sample Treatment for Protein Identification through Peptide Mass Fingerprint

Figure 5.3 shows (i) the most common in-solution protein sample treatment for protein identification, with an estimated total time of 16 h, (ii) sample treatment accelerated by ultrasound using buffered solutions, with an estimated total time of 25 min, and (iii) sample treatment accelerated by ultrasound using a salt-free solution of water/ACN, with an estimated total time of 8 min. For simplicity, we consider the sample handling to be carried out in four main steps: (i) protein denaturation, (ii) protein reduction, (iii) protein alkylation and (iv) protein digestion.

5.6.1
In-Solution Protein Denaturation

Protein denaturation is one of the most important steps in protein identification by PMF or PFF. This step directly affects the subsequent steps of reduction and alkylation, in such a way that the better the protein denaturation the better the subsequent reduction and alkylation. As we will see below, the denaturation process directly affects the results on protein coverage and number of peptides identified. This is especially true for the "clean" procedure (see below and Section 5.6.4).

Most protocols for protein denaturation consist of protein solubilization in a solution containing 6–8 M urea. However, thermal denaturation by boiling the sample, detergents such as SDS, acid-labile surfactants or mixed aqueous/organic solvents can also be used [6]. In this regard, the most important trend nowadays is linked to the need to avoid high concentrations of chaotropic agents or salts during sample treatment. As shown in the general scheme of Figure 5.3, urea and/or thiourea is used for protein denaturation, and a buffer, generally NH_4HCO_3, is also used to maintain trypsin at an adequate pH for its best activity. High salt or chaotropic agent concentrations, however, are a problem in MS determinations. Thus, measurements carried out in the MALDI ionization mode are hindered if high salt concentrations or chaotropic agents, such as SDS, are present in the sample. Such substances interfere with the correct matrix/sample crystallization process, affecting also the absorption of laser energy in the ionization mode. For measurements done with the ES ionization mode, high salt concentrations can block the capillary of the electrospray, also hindering the ionization process. Hence, to avoid interferences, a procedure called "desalting" always needs to be carried out before proceeding with the MS measurements. Generally, the desalting is performed in mini-columns filled with C_{18} beads, where peptides are retained, allowing the subsequent cleaning of the sample from salt and other substances with water. Once the sample has been cleaned, the peptides are eluted using a mixed aqueous solution of an organic solvent, generally ACN. Commercial tips with immobilized C_{18} or other types of substances are available [41]. The desalting protocol adds more steps to the method for protein identification by PMF or PFF. In addition, a percentage of peptides remains retained in the C_{18} beads. Therefore, some attempts have been reported that avoid salts and chaotropic agents; approaches that employ salts are known as "non clean" and ones that avoid them are "clean" procedures. Clean procedures are based on the solubilization/denaturation of proteins in mixed aqueous/organic solvents.

5.6.2
In-Solution Protein Reduction and Alkylation

As it can be seen in Figure 5.3, protein reduction and protein alkylation steps take 60 and 45 min, respectively, in the non-accelerated protocol. The acceleration of these steps has been studied with two ultrasonic devices, the sonoreactor and the ultrasonic probe [31]. Different times were studied: 60 min without ultrasonic energy, and 5 and 2 min with ultrasonic energy for the reduction step, and 45 min without ultrasonic energy for the alkylation step, and 5 and 2 min with ultrasonic energy. With 2 min sonication time, protein identification was not possible, with neither SR nor UP. When both steps were accelerated using 5 min sonication time, protein identification was possible, allowing the treatment time to be reduced from 60 to 5 min and from 45 to 5 min for reduction and alkylation, respectively. The acceleration of both steps was achieved with either the SR or the UP.

The possibility of avoiding cleaning steps, such as desalting procedures with ZipTips, by using ultrasonic energy and reducing the reagent concentrations by a factor of 10 has also been studied, but protein identification was not possible [31]. These results further confirm the hypothesis that ultrasonic energy does not change the mechanism of enzymatic trypsin cleavage; instead, it only enhances the enzymatic activity, probably by increasing diffusion rates and bulk temperature as a consequence of cavitation phenomena.

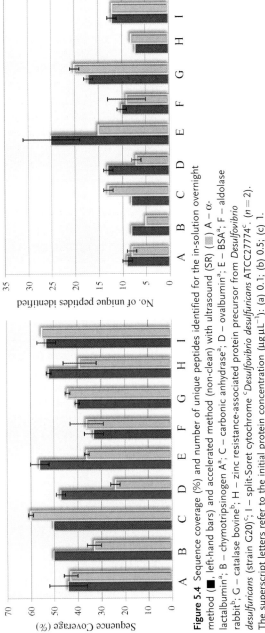

Figure 5.4 Sequence coverage (%) and number of unique peptides identified for the in-solution overnight method (■, left-hand bars) and accelerated method (non-clean) with ultrasound (SR) (▨) A – α-lactalbumin[a]; B – chymotripsinogen A[a]; C – carbonic anhydrase[a]; D – ovalbumin[a]; E – BSA[a]; F – aldolase rabbit[b]; G – catalase bovine[b]; H – zinc resistance-associated protein precursor from *Desulfovibrio desulfuricans* (strain G20)[c]; I – split-Soret cytochrome *Desulfovibrio desulfuricans* ATCC27774[c]. (n = 2). The superscript letters refer to the initial protein concentration (μg μL^{-1}): (a) 0.1; (b) 0.5; (c) 1.

5.6.3
In-Solution Protein Digestion

The traditional sample treatment for in-solution protein digestion is commonly done overnight, taking as long as 16 h. The SR, UP and UB were tested to study their ability to speed up this process. A time of 5 min was chosen for the first two ultrasonic devices while 15 min was used with the UB, due to its low intensity of sonication, as explained in Chapter 1. Unambiguous protein identification was obtained when the digestion process was accelerated with the sonoreactor and ultrasonic probe devices. Figure 5.4 presents data comparing the classic versus ultrasonic accelerated protocol using the SR. As can be seen, both methods provide similar results. Notably, the background obtained in the spectra when using MALDI-TOF-MS for protein identification was higher for the method using UP than for that using SR. When the UB was used to speed up the digestion process, protein identification was not possible, even for 15 min sonication time. These data could suggest that to speed up the kinetic process for enzymatic protein digestion there is a minimum threshold of ultrasonic cavitation efficiency required, which an ultrasonic bath cannot provide.

Regarding the solvent used to perform the digestion, both water and buffered solutions with ammonium bicarbonate or potassium phosphate have been assayed [26]. Although some positive results have been achieved with proteins dissolved in water, as a general rule, buffered solutions are recommended [31].

As far as the influence of sonication time and sonication amplitude in the acceleration process of each step is concerned, it seems that a complex relation between the type of enzyme, the type of substrate, the ultrasonic intensity and the time of sonication influences the efficiency of the whole protocol. This can be seen with two examples. Studies done with an ultrasonic probe revealed that for the protein BSA there was no large differences in the identification process developing the enzymatic digestion in the time range 60–240 s at sonication amplitudes of 10, 25 or 50%. Previously, reduction and alkylation were carried out in 5 min at 25% sonication amplitude [38]. Nevertheless, although for the case of α-lactalbumin the same conclusion was obtained, it was necessary to increase the sonication time for reduction and alkylation from 5 to 10 min at 25% sonication amplitude. Once again this finding highlights the importance of performing correctly the steps prior to the digestion process. The same work emphasized no differences in performing the different steps using two sonication frequencies of 30 or 40 kHz [38].

5.6.4
Clean In-Solution Protein Digestion

As noted several times in this chapter, a major drawback of the PMF methodology lies in the chemical denaturants, such as urea or SDS, employed to increase digestion efficiency. Such substances make the sample handling more intense, since time consuming desalting/cleaning procedures must be implemented before mass spectrometry analysis. Such procedures are needed to avoid problems during the crystallization process in MALDI and during the ionization process in MALDI or ESI.

To overcome these drawbacks, alternative methods have been developed to denature proteins prior to in-solution digestion, including thermal denaturation [42], digestion in organic solvents [24] and the use of microwave irradiation [25].

Focusing on the simplest approach, dissolution of the proteins in organic solvents, the use of ultrasonic energy needs to be carefully considered in terms of device used. For example, while the SR can be used with pure acetonitrile, this solvent rapidly evaporates when an UP is used [30, 38].

For some proteins, and depending on the protein–acetonitrile interactions, SR can be used to dissolve the proteins in pure acetonitrile. Such was the case for α-lactalbumin. Once dissolved, the protein was identified, with a sonication time of 5 min for each of the following steps: alkylation reduction and digestion. The protein coverage and the number of peptides identified were the same as for a fast procedure using urea and ultrasonication [38] but the background for the MALDI spectra was lower.

A clean approach with wide application is the one represented by dissolution of the proteins in a mixture of water : acetonitrile (1 : 1 v/v). A higher number of proteins can be solubilized than in pure acetonitrile. In addition, the procedure can be expanded to the UP, since the mixture does not volatilize as much as pure acetonitrile. With this method, either with UP or SR, protein identification can be performed easily. It is necessary only to ensure a good denaturation process, by first dissolving the proteins in buffered water, then adding the ACN and performing sonication for 1 min. This ensures a good protein denaturation and facilitates the subsequent steps of reduction and alkylation, which can be done in as little as 1 min (each step). The ultrasonic digestion step must be performed for 5 min. This fast clean method performs well for the SR or the UP. In addition, the results obtained for low protein concentrations are better than those from the method using urea and ZipTips – both the number of peptides and the sequence coverage obtained are better [38]. Furthermore, the MALDI spectra obtained with the clean protocol always have a better signal-to-noise ratio than those obtained with the urea method.

Although different trials have been carried out to simplify the number of steps in the clean method, the results seem to suggest that further research is needed to determine whether reduction and alkylation can either be omitted or simply achieved in a single step. The preliminary results seem to suggest that reduction and alkylation can be realized in a single step for BSA but not for α-lactalbumin. However, no research has been carried out to test the influence of sonication time and sonication amplitude on this matter [38].

5.7 Conclusion

Sample treatment for protein identification by peptide mass fingerprint (PMF) has been drastically changed from a labor-intensive and time-consuming method to a easier and faster one. The new approach makes use of ultrasonic energy to accelerate the four main handling steps: (i) protein denaturation, (ii) protein reduction,

(iii) protein alkylation and (iv) protein digestion . A typical in-gel approach is reduced from 24 hours to 25 min while a typical in-solution approach can be reduced from 12 to 8 min.

References

1 Hale, J.E., Gelfanova, V., Ludwig, J.R. and Knierman, M.D. (2003) *Briefings in Functional Genomics & Proteomics*, **2**, 185–193.
2 Gao, J., Garulacan, L.A., Storm, S.M. et al. (2005) *Methods*, **35**, 291–302.
3 Wetmore, B.A. and Merrick, D.A. (2004) *Toxicologic Pathology*, **32**, 619–642.
4 Steel, L.F., Haab, B.B. and Hanash, S.M. (2005) *Journal of Chromatography B*, **815**, 275–284.
5 Smejkal, G.B. and Lararev, A. (2006) *Separation Methods in Proteomics*, Taylor & Francis.
6 Lopez-Ferrer, D., Cañas, B., Vazquez, J. et al. (2006) *Trends in Analytical Chemistry*, **25**, 996–1005.
7 MASCOT search engine, available at http://www.matrixscience.com/search_form_select.html, last accessed 18 September 2008.
8 Prospector search engine, available at http://prospector.ucsf.edu/, last accessed 18 September 2008.
9 Miller, I., Crawford, J. and Gianazza, E. (2006) *Proteomics*, **6**, 5385–5408.
10 Link, A.J., Eng, J., Schieltz, D.M.,et al. (1999) *Nature Biotechnology*, **17**, 676.
11 Gorg, A., Weiss, W. and Dunn, M.J. (2004) *Proteomics*, **4**, 3665.
12 Rabilloud, T. (1008) *Electrophoresis*, **19**, 758.
13 Bjellqvist, B., Ek, K., Righetti, P.G.,et al. (1982) *Journal of Biochemical and Biophysical Methods*, **6**, 317.
14 Lauber, W.M., Carroll, J.A., Dufield, D.R. et al. (2001) *Electrophoresis*, **22**, 906.
15 Galesio, M., Vieira, D.V., Rial-Otero, R. et al. (2008) *Journal of Proteome Research*, **7**, 2097.

16 Adams, L.D. and Weaver, K.M. (1990) *Applied and Theoretical Electrophoresis*, **1**, 279.
17 Dunn, M.J. (2004) *Methods in molecular biology (Clifton, N.J.)*, **244**, 339.
18 Retamal, C.A., Thiebaut, P. and Alves, E.W. (1999) *Analytical Biochemistry*, **268**, 15.
19 Cordeiro, F.M., Carreira, R.J., Rial-otero, R. et al. (2007) *Rapid Communications in Mass Spectrometry*, **21**, 3269–3278.
20 Capelo, J.L. unpublished results.
21 Havlis, J., Thomas, H., Sebela, M. and Shevchenko, A. (2003) *Analytical Chemistry*, **75**, 1300–1306.
22 Slysz, W.G. and Schriemer, D.G. (2005) *Analytical Chemistry*, **77**, 1572–1579.
23 Slysz, W.G. and Schriemer, D.G. (2003) *Rapid Communications in Mass Spectrometry*, **17**, 1044–1050.
24 Ruseel, W.K., Park, Z.Y. and Russell, D.H. (2001) *Analytical Chemistry*, **73**, 2682–2685.
25 Pramanik, B.N., Mirza, U.A., Ing, Y.H. et al. (2002) *Protein Science*, **11**, 2676–2687.
26 Lopez-Ferrer, D., Capelo, J.L. and Vazquez, J. (2005) *Journal of Proteome Research*, **4**, 1569–1574.
27 Carreira, R.J., Cordeiro, F.M., Moro, A.J. et al. (2007) *Journal of Chromatography. A*, **1153**, 291–299.
28 Rial-Otero, R., Carreira, R.J., Cordeiro, F.M. et al. (2007) *Journal of Chromatography. A*, **1166**, 101–107.
29 Rial-Otero, R., Carreira, R.J., Cordeiro, F.M. et al. (2007) *Journal of Proteome Research*, **6**, 909–912.
30 Santos, H.M., Mota, C., Lodeiro, C. et al. Talanta, accepted.
31 Santos, H.M., Rial-Otero, R., Fernandes, L. et al. (2007) *Journal of Proteome Research*, **6**, 3393–3399.

32 Postema, M., Wamel, A., Lancee, C.T. and Jong, N. (2004) *Ultrasound in Medicine and Biology*, **30**, 827–840.

33 Wibetoe, G., Takuwa, D.T., Lund, W. and Sawula, G. (1999) *Fresenius' Journal of Analytical Chemistry*, **363**, 46–54.

34 Hielsher Ultrasound technology; http://www.Hielsher.com/ultrasonic/utr2_p.htm (last accessed March 30, 2008).

35 Kanthale, P.M., Gogate, P.R. *et al.* (2003) *Ultrasonics Sonochemistry*, **10**, 331–335.

36 Viennet, R., Ligier, V., Hihn, J.Y. *et al.* (2004) *Ultrasonics Sonochemistry*, **11**, 125–129.

37 Capelo, J.L. unpublished Results.

38 H.M. Santos, C. Lodeiro, I. Isaac, J.L. Capelo, submitted.

39 Luque-Garcia, J.L. and Neubert, T.A. (2007) *Journal of Chromatography. A*, **1153**, 259–276.

40 Wu, C.C. and Yates, J.R. (2003) *Nature Biotechnology*, **21**, 262–267.

41 http://www.millipore.com/techpublications/tech1/tn072, last accessed April 11, 2008.

42 Park, Z.Y. and Russels, D.H. (2000) *Analytical Chemistry*, **72**, 2667–2670.

6
Beyond Analytical Chemistry
Carlos Lodeiro and José-Luis Capelo-Martínez

6.1
Introduction

Beyond analytical chemistry, the applications of ultrasound (US) have long been known in industry [1], medicine and biomedical sciences [2] and in applied research [3]. The number of applications appearing in reviews of the use of ultrasound energy has in recent years increased almost exponentially. Organic and inorganic synthetic procedures gain a strong advantage on using US. Cavitation (see Chapter 1 for details) in a liquid solution provides a source of "green" energy that can be applied to enhance the rates of chemical reactions and to shorten time-consuming procedures in chemical synthesis. This physical phenomenon also has great importance in the control of reactivity. Previous chapters have demonstrated the application of US in analytical chemistry. This chapter gives an overview of its use in many chemical reactions undertaken to obtain new solid nanomaterials, polymers, new organic materials, green technologies, asymmetric synthesis and in supramolecular chemistry.

6.2
Sonochemistry for Organic Synthesis

While many organic chemists are well aware of the use of microwave, MW, assisted-reactions in organic synthesis [4] a great number are unaware of the application of US irradiation [5].

The first step in exploring the use of sonochemistry in a chemical synthesis is to find an adequate source of ultrasound energy. In most laboratories two types of ultrasound equipment are generally available: the ultrasonic bath (Figure 6.1A) and the ultrasound horn or probe system (Figure 6.1B) (see Chapter 1 for further details) [6].

The most readily available and cheapest way of applying ultrasound is the cleaning bath; the reaction vessel is immersed in the bath. The amount of ultrasonic energy

Ultrasound in Chemistry: Analytical Applications. Edited by José-Luis Capelo-Martínez
Copyright © 2009 WILEY-VCH Verlag GmbH & Co. KGaA, Weinheim
ISBN: 978-3-527-31934-3

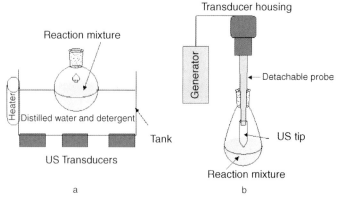

Figure 6.1 (a) Ultrasound cleaning bath and (b) ultrasound probe system.

that reaches the chemical reaction through the vessel walls is low, and the ultrasonic intensity is normally between 1 and 5 $W\,cm^{-2}$. Depending on the ultrasonic bath type, temperature control is possible. A more expensive method of applying ultrasound is the use of an US probe. This instrument permits the application of acoustic energy directly into the reaction, with the maximum ultrasonic intensities obtained being several hundred $W\,cm^{-2}$ [5].

The main drawback of the ultrasonic probe arises if the reaction needs to be performed under an inert atmosphere (Ar), under reflux and/or with pressure control. In such cases, special, expensive seals are needed to connect the horn/probe to the vessel.

Regardless of the system used to deliver ultrasonic energy, a huge number of applications concerning chemical synthesis and catalysis can be found in the literature. Some representative examples are described below.

A mild, convenient and improved method for the preparation of imines by ultrasonic irradiation has been developed by Guzen et al. (Figure 6.2) [7]. The following advantages were mentioned: simple execution, yields of 85–99% and 5 to 10 min reaction time. In addition, waste production and energy consumption were considerably reduced.

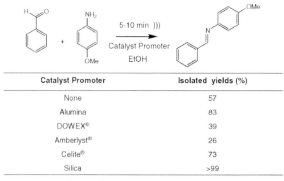

Catalyst Promoter	Isolated yields (%)
None	57
Alumina	83
DOWEX®	39
Amberlyst®	26
Celite®	73
Silica	>99

Figure 6.2 Synthesis of imines under ultrasound irradiation using solid catalysts.

Mechanical stirring and ultrasonic energy have been used and compared for the synthesis of micro-sized irregular polyanilines by dropwise addition of the ammonium persulfate solution into the aniline [8]. During the early stages of polymerization, the polymers formed in both the mechanically stirred and ultrasonicated systems are nanofibers. However, as the polymerization progresses, the primary nanofibers grow and agglomerate into irregular particles in the mechanically stirred system, while in the case of US such growth and agglomeration are avoided, thereby preserving the nanofibers in the final product. The US had no significant effect on the chemical structure of the polyaniline.

Owing to problems of selectivity, efficiency and purification, the monosubstitution of cyclodextrin derivatives is a challenge. Trotta et al. [9] have reported that such reactions are improved when carried out under an ultrasonic field (Figure 6.3). Not only were the yields higher, but reaction times were dramatically reduced to a few minutes.

The synthesis of 2-amino-1,4-naphthoquinones has become an important issue due to the antitumor and antimalarial properties of these compounds. In addition they can be used as dyes in industrial applications. The general strategies followed for their synthesis involve strong acidic conditions, the production of many by-products, low yields and intensive handling. Liu et al. [10] have reported a protocol for the synthesis of 2-amino-1,4-naphthoquinones under ultrasonic irradiation at room temperature (Figure 6.4). The method employs cheap, nontoxic molecular iodine as the catalyst and involves simple manipulation in comparison with other methodologies.

Recent advances in synthesis and catalysis have stressed the useful combination of US energy and microwave energy, MW [11]. To accomplish this the ultrasonic probe should be made of a material suitable for use in a microwave field. Obviously, probes made of metals, such as titanium, are forbidden. Nevertheless, probes of inert materials such as quartz, Pyrex, ceramic or even polyether ether ketone (PEEK) or PTFE (polytetrafluoroethylene) have been used to overcome this problem. Some examples of the simultaneous application of both US and MW are given below.

The hydrazinolysis of methyl salicylate performed under reflux for 9 h, US + reflux for 1.5 h, MW for 18 min and MW + US for 40 s afforded reaction yields of 73, 79, 80 and 84%, respectively (Figure 6.5) [12].

The classical Williamson ether synthesis involving phenols with aryl or benzyl chlorides has been accelerated from 16 h with reflux to 1–2 min using MW + US. In addition the reaction yields were increased from modest to near 80%. Neither MW nor US alone were able to cause the same effects [13].

Mannich-like reactions in aqueous and non-aqueous media have been accelerated, through MW + US, 30- and 140-fold, respectively, when compared to MW and US alone (Figure 6.6) [14].

Finally, both Suzuki and homocoupling and cross-coupling reactions involving boronic acids and aryl halides also benefit from speed and high reaction yields when performed under the combined effects of MW and US (Table 6.1) [15].

The chemistry of macrocyclic polyamines has undergone considerable development in the last 30 years. Denat et al. [16] have developed more convenient methods for the synthesis of tetraazacycloalkanes. Figure 6.7 illustrates the ultrasound

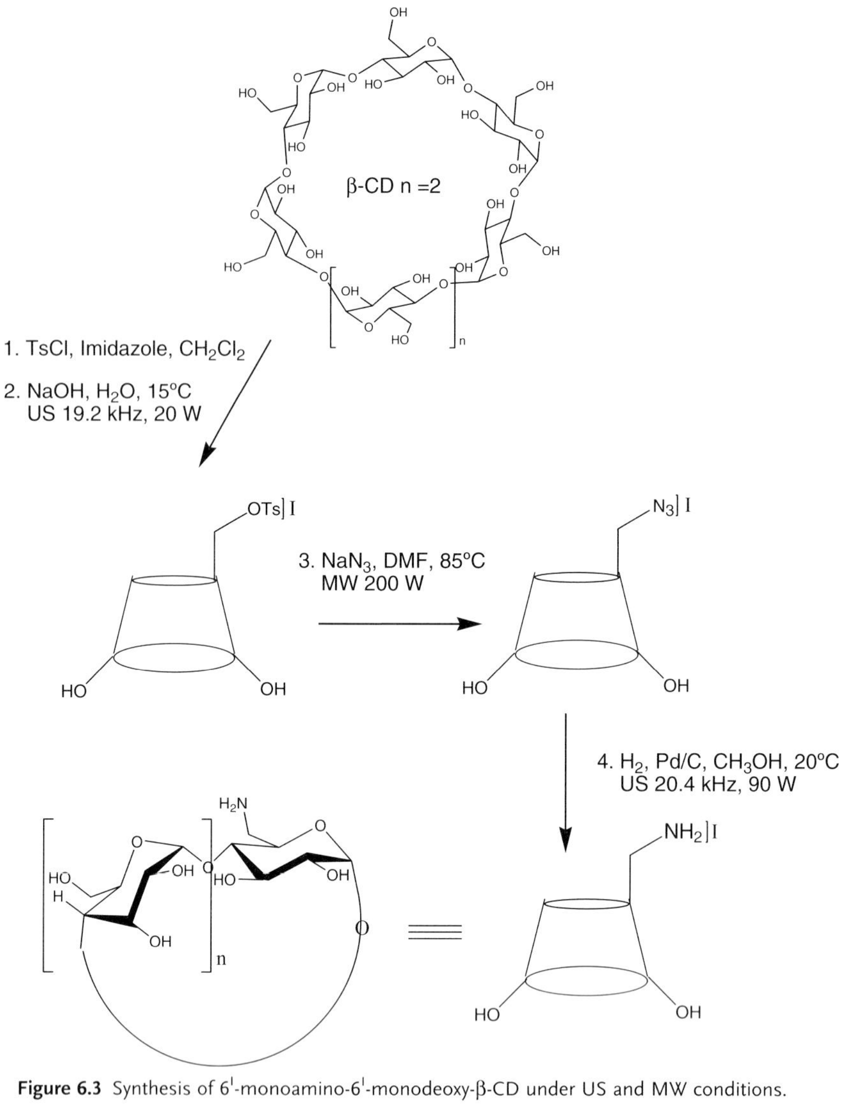

Figure 6.3 Synthesis of 6¹-monoamino-6¹-monodeoxy-β-CD under US and MW conditions.

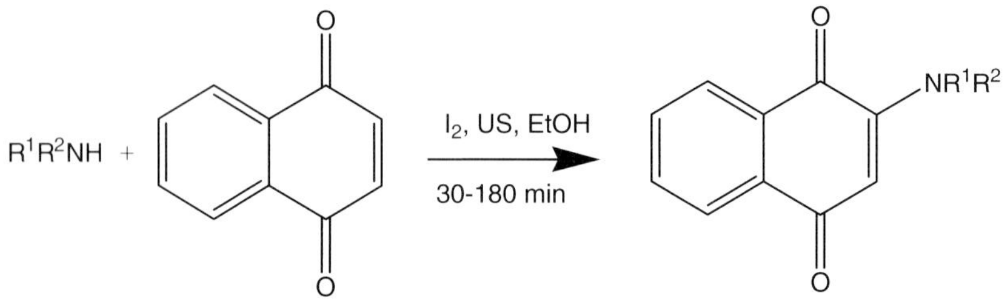

Figure 6.4 Synthetic route to 2-amino-1,4-naphthoquinones using US.

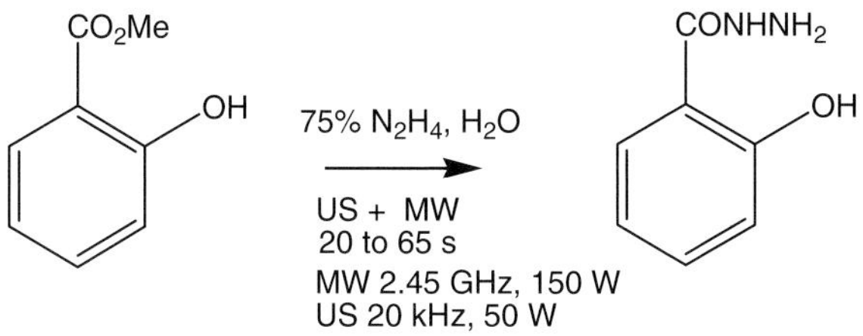

Figure 6.5 Formation of hydrazides under simultaneous MW/US irradiation.

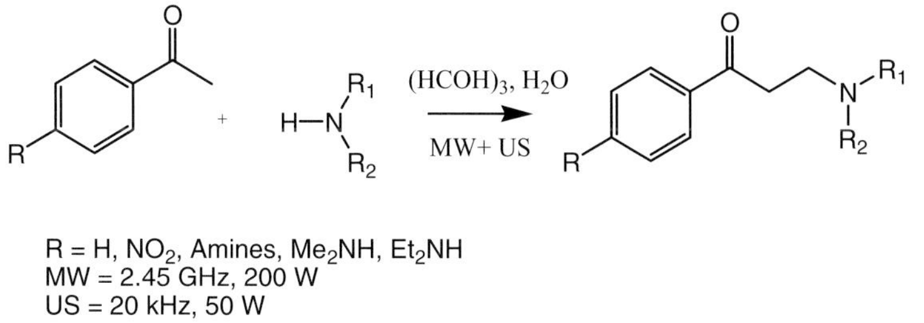

R = H, NO_2, Amines, Me_2NH, Et_2NH
MW = 2.45 GHz, 200 W
US = 20 kHz, 50 W

Figure 6.6 Enhanced Mannich reactions under US and MW irradiation in aqueous solution.

irradiation method (750 W, 20 kHz) that permits the reduction of the intermediate tetraamide derivatives in cyclan compounds.

Another important type of macrocyclic ligands, porphyrins, can be obtained in higher yields and shorter times than with conventional methods by applying

Table 6.1 Suzuki-type coupling under US and MW irradiations.

Aryl halide	Boronic acid	US yield (%)	MW yield (%)	MW/US yield (%)
3-Bromoanisole	$PhB(OH)_2$	54	64	88
2-Iodothiophene	$PhB(OH)_2$	40	37	59
4-Chloronitrobenzene	$PhB(OH)_2$	22	30	57
None	Thianthrene-1-boronic	48	55	69
None	4-tert-Butylboronic	68	74	86

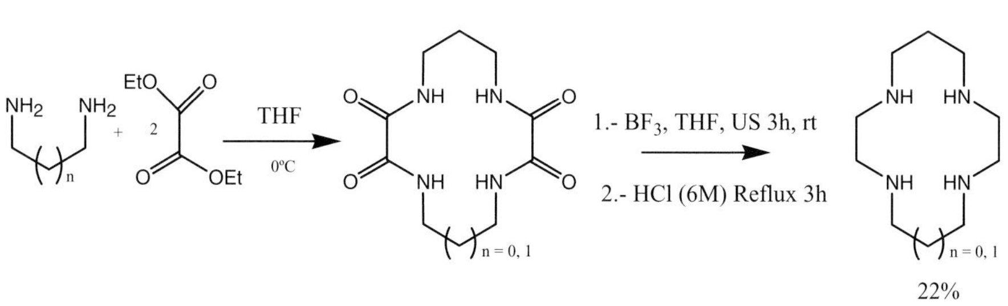

Figure 6.7 Synthesis of cyclan derivatives under US irradiation.

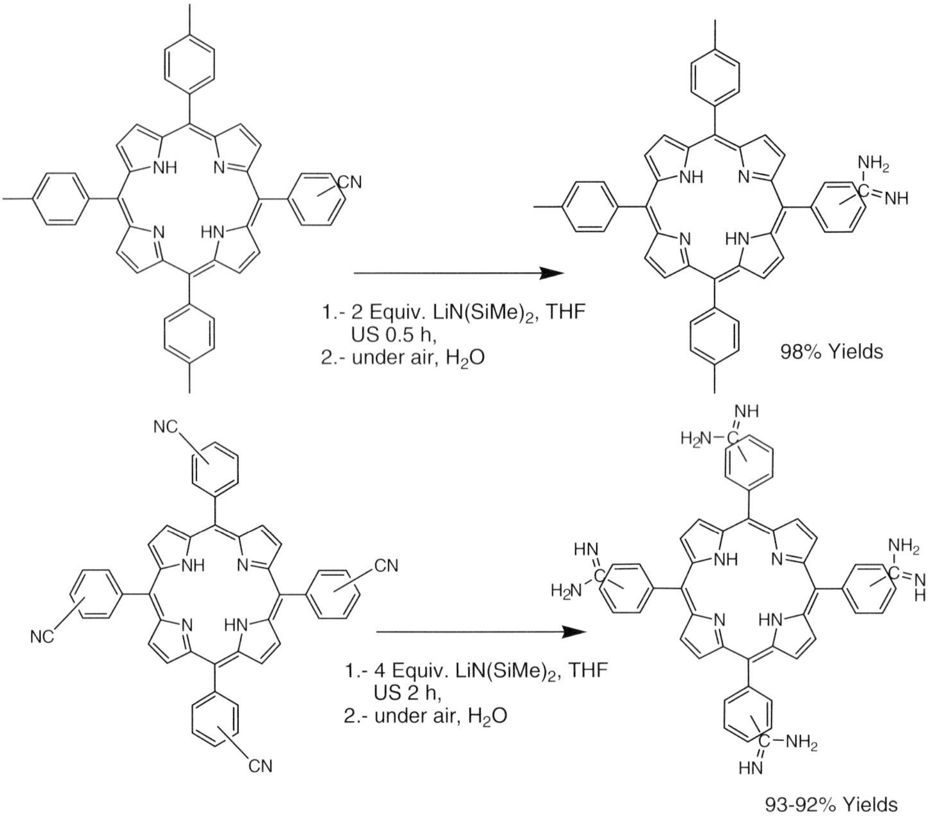

Figure 6.8 Synthesis of meso-aminophenyl monosubstituted and meso-tetrakis(aminophenyl) porphyrins.

ultrasound. Zhu et al. [17] have developed a method to obtain porphyrins via interaction of lithium amide and *meso*-cyanophenylporphyrins using an ultrasonic bath (80 W, 47 kHz) at room temperature. The reaction time was reduced to between 0.5 and 4 h from the 51 days needed by the conventional thermal method. Furthermore, the products were easily isolated by simple washing and filtration. Figure 6.8 shows the schematic reaction for mono- and tetrasubstituted porphyrin using an US irradiation bath.

Cravotto and coworkers have used high-intensity ultrasound irradiation for a novel approach to the aldol reaction in water [18]. Within 15–30 min, the acetophenone reacted with non-enolizable aldehydes, affording the aldol compound exclusively. Table 6.2 summarizes several aldol reactions of acetophenone in water under US at 18 kHz sonication frequency, 280 W of nominal power and 20 °C bulk temperature. This method offers ready access to polyols and bis-benzylidene adducts that usually can not be isolated in practical yields due to immediate elimination.

Many other important organic reactions, such as the synthesis of chalcones [19], pyrazolines [20], vitamins [21], carbohydrate compounds [22], indoles (Figure 6.9) [23],

Table 6.2 Aldol reaction of acetophenone (4 mmol) and non-enolizable aldehyde (4 mmol) in water in the presence of NaOH under US (18 kHz, 280 W, 20 °C).

Aldehyde	Product	Reaction time (min)	Yield (%)
Benzaldehyde	[Ph-CO-CH$_2$-CH(OH)-Ph]	30	73
Furfural	[Ph-CO-CH$_2$-CH(OH)-(2-furyl)]	45	59
4-Nitrobenzaldehyde	[Ph-CO-CH$_2$-CH(OH)-C$_6$H$_4$-NO$_2$]	45	78
p-Tolualdehyde	[Ph-CO-CH$_2$-CH(OH)-C$_6$H$_4$-CH$_3$]	45	46
3,4-Dimethoxybenzaldehyde	[Ph-CO-CH$_2$-CH(OH)-C$_6$H$_3$(OMe)$_2$]	45	45
4-Dimethyl aminobenzaldehyde	[Ph-CO-CH$_2$-CH(OH)-C$_6$H$_4$-NMe$_2$]	45	62
4-Methoxybenzaldehyde	[Ph-CO-CH$_2$-CH(OH)-C$_6$H$_4$-OMe]	45	37
Pyridine-3-carboxaldehyde	[Ph-CO-CH$_2$-CH(OH)-(3-pyridyl)]	45	45

(Continued)

Table 6.2 (Continued)

Aldehyde	Product	Reaction time (min)	Yield (%)
Ferrocenecarboxaldehyde	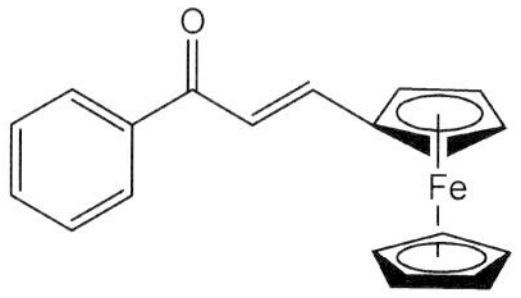	90	90
3-Nitrobenzaldehyde	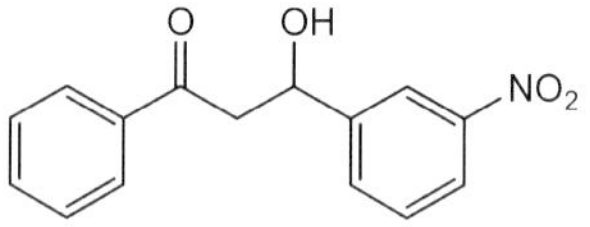	45	76

unsymmetrical reactions [24, 25] and bis(indolyl)alkanes [26], can perform better using ultrasound irradiation, with excellent activities, yields and selectivities.

Regarding organic synthesis, the utility of ultrasonic energy in the preparation of ionic liquids (IL) must be noted. Ionic liquids are a type of organic compound with a series of remarkable characteristics, such as high ionic character, virtual absence of vapor pressure, lack of flammability and ease of reuse, which make them a green alternative to common solvents used in many areas of chemistry. One of the drawbacks of ionic liquids is the long reaction times needed to synthesize them. Leveque et al. and Cravotto et al. have both reported a fast approach for the preparation of ionic liquids using ultrasonic energy that gives higher yields than the classic non-ultrasonic method [27].

The highest yields observed, 70–98%, and the short reaction times needed to complete the synthesis showed that ILs can be obtained with chloroalkanes with a considerable reduction of costs using US and the combination of MW + US irradiations. Figure 6.10 summarizes the general synthetic procedure for ILs using US and MW/US irradiation.

Finally, the effect of ultrasound on photochemical reactions has also been studied. Toma et al. [28] have concluded that the sonication process quenches long-live excited states in several molecules, because the intense streaming enhances strongly the

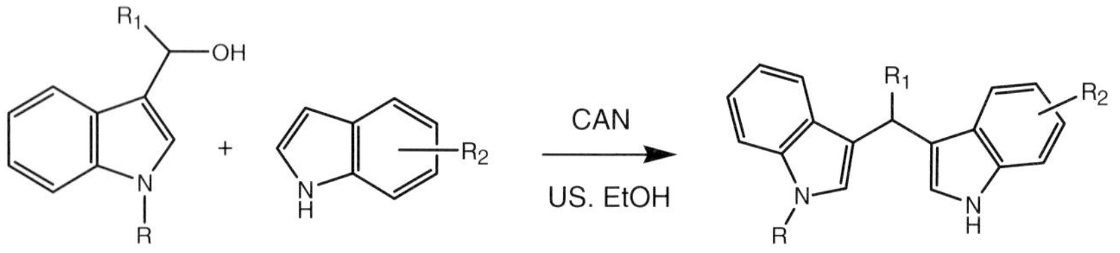

Figure 6.9 Synthesis of unsymmetrical bis(indolyl)alkanes under ultrasonic irradiation (CAN = cerium ammonium nitrate).

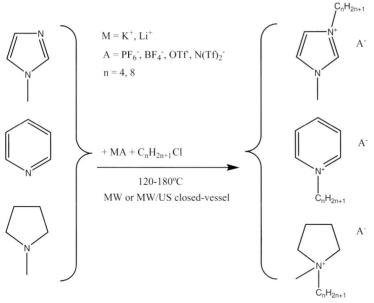

Figure 6.10 General synthetic route for ionic liquids under ultrasonic irradiation or combined MW with US in closed vessels.

probability of collisional deactivations. However, this application is still too narrow to permit decisive conclusions regarding photochemical and sonochemical effects.

6.3
Ultrasonic Enhanced Synthesis of Inorganic Nanomaterials

Fewer applications have been reported on classical inorganic chemistry, but ultrasonic irradiation has recently been applied to aspects such as catalytic reduction methodologies [29] or switching sol–gel reactions [28]. Toraishi et al. [30] have reported an enhancement of platinum black catalysis under sonochemical effects for the reduction of uranium(VI) to uranium(IV), which represents further progress in the applications of sonochemistry to the production of metal nanoparticles. Paulusse et al. [29] have reported the use of ultrasound to switch reversibly the sol–gel transition in Rh(I) and Ir(I) metal complexes. The authors propose that sonication of the gels induces a ligand exchange, which changes the topology without changing the coordination chemistry.

Complex nanoscaled electronic or optoelectronic devices are fabricated through functionalized building blocks such as nanoparticles, nanotubes and nanoplates, which can be used for different purposes in the making of nanomaterials for use, for example, in information storage or superconduction [31, 32].

Most conventional methods for the growth of nanoparticles fail in the fabrication of ring and other structures due to the difficulties inherent to this type of shape.

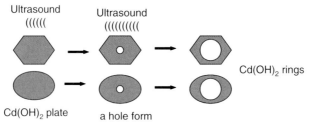

Figure 6.11 Fabrication process of Cd(OH)$_2$ nanorings by US irradiation.

During cavitation, the drastic conditions of pressure and temperature generated (>20 MPa and >5000 K) can be used for the growth of nanostructures from 0D nanoparticles, 1D nanotubes (nanorods and nanowires) and 2D nanoplates [33]. Some approaches making use of ultrasonic energy for the preparation of nanomaterials are given below.

Miao et al. [34] have reported a method for the sonochemical fabrication of single-crystalline Cd(OH)$_2$ nanorings in aqueous solution (Figure 6.11). As can be seen, through the application of US a hole gradually develops at the center of each individual nanoplate, giving rise to the Ca(OH)$_2$ nanorings.

Luminescent nanoparticles have been explored as fluorescent probes for fluorescence imaging and bioanalysis, due to their high quantum yield and multifunctional groups that provide affinity sites for the binding of biomolecules. Pan et al. [35] have proven that US can be successfully used in the synthesis of fluorescent probes through the synthesis of highly luminescent zinc(II)-bis(8-hydroxyquinoline) (Znq$_2$) complex nanorods. 8-Hydroxyquinoline reacts with Zn^{2+} in the core of microemulsions and the hydrophilic head groups of surfactant molecules, thus creating a liquid–heterogeneous system after nucleation of the Znq$_2$ nuclei. The collapse of bubbles caused by cavitation produces microjets of liquids and shockwaves that drive the solid particles to high velocities, leading to interparticle collisions. The ultimate effect is to push Znq$_2$ nanoparticles toward each other, forming the nanorods. Other systems such as electromagnetic stirring have also been used for comparative purposes, but only nanosized particles were obtained.

Gold and other noble metal nanoparticles have been intensively investigated due to their applicability to optics, electronics, catalysis and so on [36]. Jin et al. [37] have reported an environmentally friendly sonochemical method for the preparation of gold nanoparticles, GNPs, with small particle size and uniform dispersion. They demonstrated that US accelerates the formation of GNPs and helps to disperse the nanoparticles. For example, when 12 h of ultrasonication was substituted by 96 h vigorous stirring the formation of GNPs was not verified.

Chalcogenides ($E^{2-} = S^{2-}$, Se^{2-} and Te^{2-}) are materials with interesting semiconductive properties, with applications in fields such as nonlinear optics, detectors, photorefractive devices, photovoltaic solar cells and optical storage media. The mechanism proposed for the synthesis of the chalcogenides assigns as the rate-determining step the dissolution of the chalcogenide in the solvent. In this sense, US was found to have two main effects. One is favoring the dissolution of chalcogens and the formation of E^{2-}, thus accelerating the reaction. The second is to prevent aggregation of the resulting nanoparticles [38, 39].

Titanium dioxide is well known as a large-bandgap semiconductor with high photocatalytic activity. Mesoporous TiO_2 has attracted much attention because of its high surface area and large uniform pores, which are of great importance in catalysis and solar cell applications. Conventional synthesis of mesoporous TiO_2 needs several days or longer but Yu *et al.* [40] have reported a rapid synthesis of mesoporous TiO_2 with high photocatalytic activity by US-induced agglomeration. The product was obtained without a template. The authors claimed that under the effects of an ultrasonic field, produced under high intensity ultrasound irradiation, it is possible to control the condensation and agglomeration of monodispersed sol particles, leading to the formation of mesoporous structures with a narrow size distribution. They have called this phenomenon ultrasound-induced agglomeration.

6.4
Sonochemistry Applied to Polymer Science

6.4.1
Introduction to Polymers

A polymer is a substance composed of repeating structural units, also called monomers, generally connected by covalent bonds. The word means "many parts" as it comes from the ancient Greek word "polu-meros" (οπλυ-μερος). Examples of well-known polymers are plastics, proteins and natural polymers, such as polypropylene, hair or skin, and cellulose, respectively.

Synthetic polymers find application in nearly every industry and area of life. They are used as adhesives, lubricants or structural components for many different materials, from computers to satellites. The most recent applications are found in the bioscience area, from artificial components for the human body to drug delivery.

The way in which monomers are distributed to form the structure of the polymer dictates the polymer's properties, such as solubility, durability, resistibility, crystallinity or tensile strength.

The characterization of a polymer is mandatory to define its properties, since the monomers are distributed statistically in chains of variable lengths, which gives rise to differences in physical and chemical characteristics.

Different analytical techniques can be used to define a polymer, including wide Fourier-transform infrared spectroscopy, wide-angle X-ray scattering, small-angle neutron scattering, gel permeation chromatography, Raman spectroscopy, small-angle X-ray scattering, nuclear magnetic resonance, poly-dispersity and mass spectrometry-based techniques.

Ultrasonication has become a tool in polymer research with three different aims. First, ultrasonication has long been used in the synthesis of polymers, as an initiator or to obtain a homogeneous distribution of the monomers. Second, it has also been used to study polymer degradation mechanisms. Third, a recent approach deals with the utilization of ultrasonication for fast sample treatment prior to the characterization of polymers by mass spectrometry-based techniques.

The next sections deal with the above-mentioned applications of US energy in research dealing with the analytical characterization of polymers.

6.4.2
Ultrasonication in Sample Treatment for Polymer Characterization

6.4.2.1 Introduction to Polymer Characterization

Several parameters are used to characterize a polymer [41]. The most important are the number-average molecular weight (M_n), the weight-average molecular weight (M_w) and the mass of repeat units, also called the polydispersity.

The weight-average molecular weight is a way of describing the molecular weight of a polymer. A problem arises because it is not possible to obtain the same polymer molecule repeatedly since it comes in different sizes, for example, different chain lengths for linear polymers. For this reason, an average of the weight is taken for a series of samples from the same polymer. This value is calculated through the following formula:

$$\bar{M}_w = \frac{\sum_i N_i M_i^2}{\sum_i N_i M_i}$$

were Ni is the number of molecules with a molecular weight of Mi. In other words, the weight-average molecular weight provides an average weight of a given polymer.

The $\bar{M}_w$ can be calculated by different analytical techniques, including small-angle neutron scattering (SANS), X-ray scattering, sedimentation velocity and mass spectrometry-based techniques.

The number-average molecular weight, $\bar{M}_n$, is another form of the molecular weight of a polymer, but in this case $\bar{M}_n$ is calculated as the arithmetic mean or average of the molecular weights of a certain number of molecules. Therefore, $\bar{M}_n$ is calculated by measuring the molecular weight of n polymer molecules, summing the weights, and dividing by n, as follows:

$$\bar{M}_n = \frac{\sum_i N_i M_i}{\sum_i N_i}$$

The number-average molecular weight of a polymer can be determined by gel permeation chromatography, by mass spectrometry-based techniques, viscosimetry and all colligative methods like vapor pressure osmometry or end-group titration.

The ratio $\bar{M}_w$ to $\bar{M}_n$ is known as the polydispersity index, PDI, and is another polymer characteristic that measures the distribution of molecular masses in a given polymer sample. The PDI always has a value greater than 1, approaching unity as the polymer chain approaches a uniform chain length. Gel permeation chromatography, size exclusion chromatography or mass spectrometry-based techniques can be used to calculate the PDI.

Besides the above-mentioned features, polymer characterization involves the determination of the repeat unit structure, copolymers sequence analysis and end-group determination.

Matrix-assisted laser desorption/ionization time-of-flight mass spectrometry, MALDI, has become a routine analytical tool for the structural analysis of polymers. Some of the advantages of this technique are [42–44]: (i) absolute molecular weights of narrowly distributed polymers (polydispersity < 1.2) can be determined as opposed to relative molecular weights obtained by chromatographic techniques; (ii) analysis does not require polymer standards to assign molecular weights to oligomers; (iii) using submilligram amounts of sample material, the analysis can be accomplished in a few minutes; and (iv) MALDI can determine the molecular weight independently of the polymer structure . In addition, MALDI can be used in conjunction with liquid chromatography, thereby combining the advantages of both methodologies.

The above-described formulas for the calculation of $\bar{M}_w$, $\bar{M}_n$ and PDI are easily transformed in the MALDI analytical methodology, as follows:

$$\bar{M}_w = \frac{\sum_i IiMi^2}{\sum_i IiMi}$$

and:

$$\bar{M}_n = \frac{\sum_i IiMi}{\sum_i Ii}$$

where Ii is the signal intensity in the MALDI for a molecular weight of Mi, assuming that there is a linear relationship between the number of ions and signal intensity.

Several instrumental parameters that can affect the values obtained for $\bar{M}_w$ and $\bar{M}_n$ must be controlled in the MALDI analysis, namely, the detector voltage, laser energy, delay time, extraction voltage and lens voltage [41, 45, 46]. Once these parameters are under control, the main variable in determining $\bar{M}_w$ and $\bar{M}_n$ arises from the sample treatment and sample deposition on the MALDI plate.

6.4.2.2 Overview of Sample Preparation for MALDI Analysis of Polymers

To perform an analysis in MALDI, the polymer needs to be dissolved in a solution called a matrix. The matrix consists of small organic molecules that strongly absorb the laser light. The energy absorbed is transmitted to the mixture in such a way that instantaneous vaporization is induced in the area where the laser beam impinges. Then, a mixture of ionized matrix and analyte molecules called the "plume" is released into the vacuum of the MALDI system.

Many matrixes are available for polymer analysis, and one critical step of the procedure is to choose the correct matrix for the polymer to be studied. An adequate matrix is selected by focusing on the backbone structure of the polymer and then searching the literature for a set of candidate matrixes, which are then used to obtain the respective MALDI spectra. Finally, the matrix giving the best spectrum is selected for further studies [47]. Hoteling *et al.* [48] have pointed out that the best matrix–polymer couple can be found using reversed-phase HPLC: when the matrix and the polymer have similar retention times it is likely that they match well for the MALDI analysis.

It was generally agreed until recently that incorporation of individual analyte molecules into the crystalline host matrix was mandatory to obtain well-defined MALDI spectra, but nowadays this seems to be less important [38].

The method of sample deposition in the MALDI target most frequently reported is the so-called "dried droplet." In this method matrix, analyte and salts are mixed and the mixture is spotted onto the MALDI target – it is possible to prepare more than 100 targets per hour [49]. However, both matrix and polymer must be soluble in the same solvent, otherwise selective crystallization of the analyte, matrix or cationizating agent can occur, giving rise to significant variations in peaks, peak intensity, resolution and mass accuracy when the laser is focused on different regions of the same spot.

Electrospray sample deposition is a method that involves spraying the matrix and analyte solutions onto the MALDI target using a high-voltage electrical field. This allows the formation of smaller crystals and, therefore, increases the reproducibility of shot-to-shot and spot-to-spot analysis [50].

Other methods for sample treatment for polymer analysis can be found in the literature and new ones appear from time to time. For instance, the solvent-free sample preparation or the wet grinding methods.

Concerning the dried droplet method, the matrix and polymers are generally mixed using vortex mixing – a method that only permits the mixing of samples one by one.

To avoid this problem different trials have been carried out to increase sample throughput by using ultrasonic energy as a way to rapidly mix polymer and matrix. It must be stressed that ultrasonication has long been studied as a way to degrade polymers and also, as we will see below, as a way to characterize polymers through ultrasonic degradation and the degradation mechanism. Consequently, US energy can only be employed for mixing samples for polymer analysis if short sonication times are used.

6.4.2.3 Ultrasonic Energy as a Tool for Fast Sample Treatment for the Characterization of Polymers by MALDI

In recent work [51], different ultrasonic devices were studied to compare their performance for fast sample preparation (mixing) of polymers and matrixes. The sonic devices chosen were the ultrasonic bath (UB), the ultrasonic probe (UP) and the sonoreactor (SR). Table 6.3 lists the conditions used, where it can be seen that a vortex was used as the standard method for sample mixing.

For the same sonication amplitude, as the sonication frequency increases so the sonication intensity decreases (Chapter 1). For this reason different frequencies of sonication were assayed. On the other hand, for the same sonication frequency, the longer the sonication time the higher the risk of polymer degradation. Consequently, the study also required the testing of different sonication times for the same frequency of sonication. Regarding the amplitude of sonication, it is well known that UBs can not produce high intensity sonication. Therefore, the amplitude chosen for this device was 100%, so as to obtain the highest sonication intensity possible. Conversely, 20%

6.4 Sonochemistry Applied to Polymer Science

Table 6.3 Conditions for the different sample treatments studied for polymer/matrix mixing under the effects of an ultrasonic field.

Device for homogenization	Sonication frequency (kHz)	Treatment time (s)	Sonication amplitude (%)
Ultrasonic bath	130	120	100
Ultrasonic bath	35	30	100
Ultrasonic probe	40	30	20
Ultrasonic probe	40	10	20
Sonoreactor	35	30	20
Sonoreactor	35	10	20
Vortex	—	60	—

sonication amplitude was chosen for the most intense devices, the UP and the SR, since it was felt that higher amplitudes could lead to polymer degradation.

Table 6.4 shows the results of this preliminary study regarding the applicability of ultrasonication to the sample treatment of polymers. The study was performed on two different polymers, polystyrene (PS) and poly(ethylene glycol) (PEG), with two different matrixes, dithranol and 2,5-dihydroxybenzoic acid (DHB). Figure 6.12 shows the chemical forms of these polymers and matrixes.

Concerning the UP, it was found that for PEG analysis using dithranol as matrix, the values obtained for $\bar{M}_w$, and $\bar{M}_n$ were statistically different from those obtained from vortex mixing, thus indicating a difference that can only be related to the sample mixing procedure. This result was independent of the sonication time used with the UP, either 10 or 30 s. Notably, the analysis performed for PEG with DHB using an UP gives results similar to those obtained with the vortex mixing and with the values reported by the manufacturer of the polymer.

Regarding the sonoreactor, the analysis of PEG with dithranol or DHB, when 30 s sonication time was used, gave results statistically different to both those obtained with the vortex mixing and the values reported by the manufacturer of the polymer. Interestingly, when the sonication time used was only 10 s no differences were found. However, this variation over so short a time range, from 10 to 30 s, indicates the low robustness of this setup for systematic use in sample treatment of polymers.

The most promising device seems to be the ultrasonic bath. With this system two different frequencies of sonication, 35 and 130 kHz, were used. The homogenization process with the latter frequency for the PS 2000 Da in DHB matrix during 120 s led to low $\bar{M}_w$ and $\bar{M}_n$ values that were statistically different from those obtained for sample homogenization using vortex mixing and from the values given by the polymer manufacturer. Despite these results, when an UB at a sonication frequency of 35 kHz was used the $\bar{M}_w$ and $\bar{M}_n$ obtained for PS and PEG in dithranol matrix were not statistically different from the ones acquired with vortex mixing or from the values recommended by the manufacturers.

This pioneering work thus suggests that the UB with a sonication frequency of 35 kHz could be used for fast, high-throughput sample treatment of polymers for their characterization. Nevertheless, this methodology needs further confirmation

Table 6.4 M_n and M_w values ± standard deviations[a] of three polymers in dithranol and DHB matrixes for each of the seven sample treatments tested.

Polymer	PS 2000				PS 10 000				PEG 1000			
	Dithranol		DHB		Dithranol				Dithranol		DHB	
Sample treatment	M_n	M_w	M_n	M_w	M_n	M_w			M_n	M_w	M_n	M_w
US bath 130 kHz, 120 s	2246 ± 24	2382 ± 29	1755 ± 71	1855 ± 75	8711 ± 47	8828 ± 41			1009 ± 8	1037 ± 8	1028 ± 1	1058 ± 1
US bath 35 kHz, 30 s	2250 ± 42	2371 ± 57	1826 ± 22	1930 ± 25	8681 ± 38	8795 ± 28			1001 ± 11	1033 ± 10	1003 ± 20	1033 ± 21
US cell disruptor 10 s	2234 ± 26	2378 ± 23	1922 ± 70	2028 ± 66	8524 ± 146	8638 ± 145			1024 ± 7	1054 ± 9	997 ± 5	1031 ± 3
US cell disruptor 30 s	2245 ± 70	2392 ± 71	1876 ± 10	1993 ± 11	8625 ± 72	8743 ± 70			1024 ± 6	1053 ± 4	1010 ± 14	1045 ± 14
Sonoreactor 10 s	2156 ± 81	2312 ± 98	1921 ± 25	2029 ± 31	8726 ± 63	8837 ± 64			1014 ± 11	1043 ± 8	1029 ± 8	1058 ± 9
Sonoreactor 30 s	2176 ± 35	2313 ± 20	1954 ± 60	2059 ± 52	8646 ± 95	8757 ± 94			1029 ± 12	1056 ± 11	1039 ± 2	1069 ± 1
Vortex	2198 ± 39	2352 ± 56	1888 ± 95	2004 ± 92	8615 ± 72	8727 ± 67			1003 ± 7	1035 ± 5	1014 ± 4	1044 ± 6
Theoretical value	2140	2250	2140	2250	8650	8900			900	940	900	940

[a] M_n and M_w values are averages of the data obtained for three samples ($n = 3$).

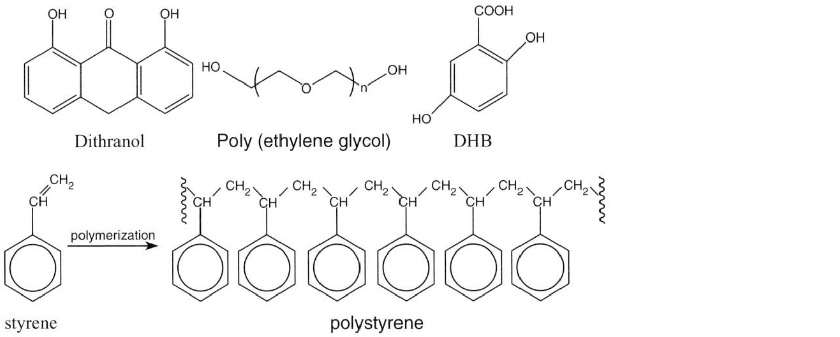

Figure 6.12 Chemical structures of dithranol, poly(ethylene glycol), DHB, styrene and polystyrene.

and to be extended to more polymers. Consequently, more work dealing with this subject is anticipated.

6.4.3
Ultrasonic-Induced Polymer Degradation for Polymer Characterization

Kawasaki et al. [52] have recently stressed the possibility of using ultrasonic fragmentation for the characterization of high molecular weight synthetic polymers using ultrasonic degradation. The key to this approach is the finding that, regardless of size, the polymer always fragments in the same way under the effects of an ultrasonic field. This was the case for PEG with average molecular masses of 2, 6, 20 and 2000 kDa. After a dedicated study using MALDI, the authors proposed the ultrasonic degradation processes shown in Figure 6.13. In conclusion, it was possible to study the structural details of ultrasonic degradation of PEG, showing five types of degradation products with different end-groups, irrespective of the initial molecular masses. Five degradation pathways involving free radical reactions are proposed: the first route is ultrasonic scission of PEG by C–O bond breaking, yielding two daughter products with different terminal radical groups (Figure 6.13). The subsequent route is termination of the ultrasonically generated daughter groups by extraction or release of a hydrogen atom. Further degradation of the resultant products yields the polymeric radical of $CH_3CH_2–OCH_2CH_2$. This radical then extracts a hydrogen atom, leading to another product.

If more mechanisms of ultrasonic polymer degradation are confirmed as being reproducible regardless of polymer size, as shown above, then it will open up a new way to characterizing high molecular weight synthetic polymers using MALDI and ultrasonic energy.

6.4.4
Ultrasonication for the Preparation of Imprinted Polymers

Molecular imprinting is a method that uses highly crosslinked polymers to produce synthetic molecule receptors with a predetermined selectivity. By casting the polymer

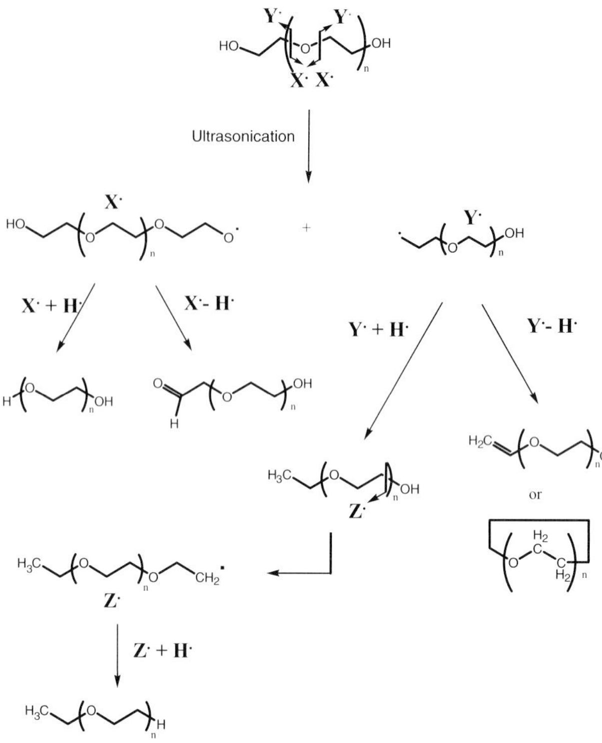

Figure 6.13 Ultrasonic degradation pathways followed by PEG polymers proposed by Kawasaki et al. [52]. The C—O bonds are broken through ultrasonic energy, giving two polymer radicals, Ẋ(–CH$_2$CH$_2$O) and Ẏ(–CH$_2$CH$_2$). The first radical, by accepting or losing one H˙ gives rise to two products. A similar process occurs for the radical Ẏ(~CH$_2$CH$_2$), which affords three different products.

in the presence of a substance that acts as template, it is possible to prepare a polymer with cavities that reflect the shape and the chemical functionality of the template. The process can be likened to the imprint created by a hand pressed in wet cement. Once the wet cement has dried, the shape of the hand remains in such a way that only the same hand or another of the same size will be "recognized" by the imprint. Figure 6.14 shows schematically the preparation of a molecularly imprinted polymer (MIP).

A MIP exhibits characteristics similar to the template from which it was created. This is very important, because when the template used is an enzyme or an antibody the MIP can be used as an alternative to those biomolecules. Applications of MIPs include, among others, different features such as antibody mimics in binding assays [53], stationary phases in liquid chromatography [54], electrochromatography [55], in solid phase extraction [56], as region- and stereospecific catalysis and enzyme mimics [57, 58] and as the basis for membranes and sensors [59, 60].

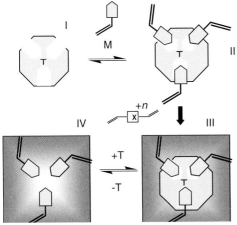

Figure 6.14 Representation of the molecular imprinting process: (I) the template (T) and the monomer (M); (II) a solution reaction between T and M; (III) the polymer is formed after the addition of a crosslinking agent and an initiator; finally, removing the template leaves, in IV, the imprinted polymer with specific functional groups.

Historically, since it was first reported as an initiator by Ostroski and Staumbaugh in 1950 [61] for the preparation of polystyrene, ultrasonication has been applied to reactions of polymerization, including homopolymerization of pure monomers, step-growth and emulsion polymerization [61–64].

Theoretically, ultrasonication could be used for the following purposes regarding the preparation of MIPs: (i) to aid in the initiation, (ii) to increase the solubility of the template and the monomers and (iii) to achieve efficient degassing throughout the polymerization. It is also expected to alter the binding sites population distribution as well as the morphology of the finished polymers.

In his work on the influence of ultrasonication in the elaboration of a MIP towards theophylline [65], Svenson pointed out that the polymers elaborated by a standard procedure and by a procedure using ultrasonication were identical in appearance and that the workup did not reveal any structural differences, as the polymers were obtained in similar yields (60–65%) after grinding and sedimentation. In fact, studies regarding morphology showed comparable surface areas and pore size distribution regardless of the mode of preparation. More interestingly, the capacity factors, binding capacity and the recognition characteristics were also similar. The advantages described by this author for MIPs made under the effects of an ultrasonic field refer to the possibility of initiating polymerization at lower temperatures than with the reference method (i.e., without ultrasonication) and to the chromatographic behavior of the sonicated polymers, which can be operated at lower back pressure, suggesting a narrower binding site distribution, although this does not necessarily means higher affinity binding sites. In any case, the result is a beneficial reduction in peak tailing. Peak tailing has been cited as one of the drawbacks of using MIPs in chromatographic approaches, when the MIPs are formed without ultrasonic energy. In addition to the advantages in peak tailing, it was also reported that the heights of the chromatographic peaks increased by 35% when the polymer was obtained with ultrasonication.

6.5
Conclusion

This chapter briefly overviews ultrasonic applications beyond analytical chemistry, although some of the applications described are directly linked with those in analytical chemistry, such as is the case for imprinting polymers or ionic liquids. The different approaches covered demonstrate that ultrasound (US) energy is a constantly growing tool in analytical chemistry and related disciplines. An improvement of classic methods, for synthesis or any other chemical field, that entails US is always worthy of note since many different aspects, from mass transfer to non-agglomeration process, can be improved and can be performed in shorter times and give higher yields.

Acknowledgments

Financial support given by the Portuguese Science Foundation and FEDER (Portugal/UE) under project POCI/QUI/55519/2004 is acknowledged.

References

1 Gronroos, A., Pirkonen, P. and Ruppert, O. (2004) *Ultrasonics Sonochemistry*, **11**, 9–12. Albu, S., Joyce, E., Paniwnyk, L. et al. (2004) *Ultrasonics Sonochemistry*, **11**, 261–265. Jiranek, V., Grbin, P., Yap, A. et al. (2008) *Biotechnology Letters*, **30**, 1–6. Sivakumar, V., Verma, V.R., Swaminathan, G. and Rao, P.G. (2007) *Journal of Scientific and Industrial Research*, **66**, 545–549. Muthurkumaran, S., Kentish, S.E., Stevens, G.W. and Ashokkumar, M. (2006) *Reviews in Chemical Engineering*, **22**, 155–194. Scanlon, M.G. (2004) *Food Research International*, **37**, 535–536.

2 Diederich, C.J., Stafford, R.J., Burdette, W.H. et al. (2004) *Medical Physics*, **31**, 405–413. Melodelima, D., Theillere, J.Y. and Cathingnol, D. (2004) *Ultrasound in Medicine and Biology*, **30**, 103–111. Yu, T.H., Wang, Z.B. and Mason, T.J. (2004) *Ultrasonics Sonochemistry*, **11**, 95–103. Brujan, E.A. (2004) *Ultrasound in Medicine and Biology*, **30**, 381–387. Liu, Y., Yang, H. and Sakanishi, A. (2006) *Biotechnology Advances*, **24**, 1–16. Zhang, D., Gong, X.F., Liu, J.H. et al. (2000) *Ultrasound in Medicine and Biology*, **26**, 347–351. Avivi, S., Felner, I., Novik, I. and Gedanken, A. (2001) *Biochimica et Biophysica Acta*, **1527**, 123–129. Hoogland, R. (1986) *Ultrasound Therapy*, Enraf Nonius, Delft, Holland.

3 Salleh-Mack, S.Z. and Roberts, J.S. (2007) *Ultrasonics Sonochemistry*, **14**, 323–329. Ruecroft, G., Hipkiss, D., Ly, T. et al. (2005) *Organic Process Research & Development*, **9**, 923–932. Mason, T.J., Collings, A. and

Sumel, A. (2004) *Ultrasonics Sonochemistry*, **11**, 205–210. Mason, T.J. (2003) *Ultrasonics Sonochemistry*, **10**, 175–179.

4 Tanaka, K. (2003) *Solvent-free Organic Synthesis*, Wiley-VCH Verlag GmbH, Weinheim. Loupy, A. (2002) *Microwaves in Organic Synthesis*, Wiley-VCH Verlag, GmbH, Weinheim. Cravotto, G. and Cintas, P. (2007) *Chemistry – A European Journal*, **13**, 1902–1909.

5 Mason, T.J. (1997) *Chemical Society Reviews*, **26**, 443–451.

6 Mason, T.J. (1991) *Practical Sonochemistry. A Users Guide to Applications in Chemistry and Chemical Engineering*, Ellis Horwood Publishers, Chichester.

7 Guzen, K.P., Guarezemini, A.S., Orfao, A.T.G. et al. (2007) *Tetrahedron Letters*, **48**, 1845–1848.

8 Jing, X., Wang, Y., Wu, D. and Qiang, J. (2007) *Ultrasonics Sonochemistry*, **14**, 75–80.

9 Trotta, F., Martina, K., Ribaldo, B. et al. (2007) *Journal of Inclusion Phenomena and Molecular Recognition in Chemistry*, **57**, 3–7.

10 Liu, B. and Ji, S.-J. (2003) *Synthetic Communications*, **33**, 1777–1789.

11 Cravotto, G. and Cintas, P. (2007) *Chemistry – A European Journal*, **13**, 1902–1909.

12 Peng, Y. and Song, G. (2001) *Green Chemistry*, **3**, 302–304.

13 Peng, Y. and Song, G. (2002) *Green Chemistry*, **4**, 349–351.

14 Peng, Y. and Song, G. (2003) *Green Chemistry*, **5**, 704–706.

15 Cravotto, G., Beggiatto, M., Penoni, A. et al. (2005) *Tetrahedron Letters*, **46**, 2267–2271. Ying, L. and Liebscher, J. (2007) *Chemical Reviews*, **107**, 133–173.

16 Denat, F., Tripier, R., Boschetti, F. et al. (2006) *Arkivoc*, **4**, 212.

17 Zhu, X.-J., Wong, W.-K., Jiang, F-L. et al. (2008) *Tetrahedron Letters*, **49**, 2114–2118.

18 Cravotto, G., Demetri, A., Nano, G.M. et al. (2003) *European Journal of Organic Chemistry*, **22**, 4438–4444.

19 Durán-Valle, C.J., Fonseca, I.M., Calviño-Casilda, V. et al. (2005) *Catalysis Today*, **107–108**, 500–506. Wei, W., Qunrong, W., Liqin, D. et al. (2005) *Ultrasonics Sonochemistry*, **12**, 411–414.

20 Mallouk, S., Bougrin, K., Doua, H. et al. (2004) *Tetrahedron Letters*, **45**, 4143–4148.

21 Bonrath, W. (2003) *Ultrasonics Sonochemistry*, **10**, 55–59. Bonrath, W. (2004) *Ultrasonics Sonochemistry*, **11**, 1–4.

22 Kardos, N. and Luche, J.-L. (2001) *Carbohydrate Research*, **332**, 115–131.

23 Palimkar, S.S., Kumar, P.H., Lahoti, R.J. and Srinivasan, K.V. (2006) *Tetrahedron*, **62**, 5109–5115.

24 Polàcková, V., Huťka, M. and Toma, S. (2005) *Ultrasonics Sonochemistry*, **12**, 99–102.

25 Török, B., Balázsik, K., Felföldi, K. and Bartók, M. (2001) *Ultrasonics Sonochemistry*, **8**, 191–200.

26 Zeng, X.-F., Ji, S.-J. and Wang, S.-Y. (2005) *Tetrahedron*, **61**, 10235–10241.

27 Leveque, J.-M., Luche, J.-L., Petrier, C. et al. (2002) *Green Chemistry*, **4**, 357–360. Lévêque, J.M., Cravotto, G., Boffa, L. et al. (2007) *Synlett*, **3**, 2065–2068. Cravotto, G., Boffa, L., Lévêque, J.-M. et al. (2007) *Australian Journal of Chemistry*, **60**, 946–950. Cravotto, G., Gaudino, E.C., Boffa, L. et al. (2008) *Molecules*, **13**, 149–156.

28 Toma, S., Goplovsky, A. and Luche, J.-L. (2001) *Ultrasonics Sonochemistry*, **8**, 201–207.

29 Paulusse, J.M.J., Beek, D.J.M. and Sijbesma, R.P. (2007) *Journal of the American Chemical Society*, **129**, 2392–2397.

30 Toraishi, T., Kimura, T. and Arisaka, M. (2007) *Chemical Communications*, 240–241.

31 Zhu, J.G., Zheng, Y.F. and Prinz, G.A. (2000) *Journal of Applied Physics*, **87**, 6668–6673. Geng, J., Hou, W.H., Lu, Y.N. et al. (2005) *Inorganic Chemistry*, **44**, 8503–8509.

32 Aizpurua, J., Hanarp, P., Sutherland, D.S. et al. (2003) *Physical Review Letters*, **90**, 57401.

33 Gedanken, A. (2004) *Ultrasonics Sonochemistry*, **11**, 47–55.

34 Miao, J.J., Fu, R.-L., Zhu, J.-M. *et al.* (2006) *Chemical Communications*, 3013–3015.

35 Pan, H.-C., Liang, F.-P., Mao, C.-J. *et al.* (2007) *The Journal of Physical Chemistry. B*, **111**, 5767–5772.

36 Schmid, G. (1992) *Chemical Reviews*, **92**, 1709–1727.

37 Jin, Y., Wang, P., Yin, D. *et al.* (2007) *Colloids and Surfaces A-Physicochemical and Engineering Aspects*, **302**, 366–370.

38 Gedanken, A. (2004) *Ultrasonics Sonochemistry*, **11**, 47–55.

39 Li, Q., Ding, Y., Shao, M.W. *et al.* (2003) *Materials Research Bulletin*, **38**, 539–543.

40 Yu, J.C., Zhang, L. and Yu, J. (2002) *New Journal of Chemistry*, **26**, 416–420.

41 Montaudo, G., Samperi, F. and Montaudo, M.S. (2006) *Progress in Polymer Science*, **31**, 277–357.

42 Macha, S.F. and Limbach, P.A. (2002) *Current Opinion in Solid State & Materials Science*, **6**, 213.

43 Wu, K.J. and Odom, R.W. (1998) *Analytical Chemistry*, **70**, 456A.

44 Wetzel, S.J., Guttman, C.M., Flynn, K.M. and Filliben, J.J. (2006) *Journal of the American Society for Mass Spectrometry*, **17**, 246.

45 Wetzel, S.J., Guttman, C.M. and Girard, J.E. (2004) *International Journal of Mass Spectrometry*, **238**, 215.

46 Wetzel, S.J., Guttman, C.M. and Flynn, K.M. (2004) *Rapid Communications in Mass Spectrometry*, **18**, 1139.

47 NIST. National Institute of standards and technology, Polymers Division, http://polymers.msel.nist.gov/ (last accessed April 30, 2008).

48 Hoteling, A.J., Erb, W.J., Tyson, R. and Owens, K.G. (2004) *Analytical Chemistry*, **76**, 5157–5164.

49 Bahr, U., Deppe, A., Karas, M. *et al.* (1992) *Analytical Chemistry*, **64**, 2866–2869.

50 Wetzel, S.J., Guttman, C.M. and Flynn, K.M. (2004) *Rapid Communications in Mass Spectrometry*, **18**, 1139–1146.

51 Fernandes, L., Rial-Otero., R., Temtem, M. *et al.* (2008) *Talanta* (in press) doi: 10.1016/j.talanta.2008.07.049

52 Kawasaki, H., Takeda, Y. and Arakawa, R. (2007) *Analytical Chemistry*, **79**, 4182–4187.

53 Vlatakis, G., Andersson, L.I., Muller, R. and Mosbach, K. (1993) *Nature*, **361**, 645–647.

54 Sellergren, B. and Shea, K.J. (1993) *Journal of Chromatography. A*, **635**, 31–49.

55 Schweitz, L., Andersson, L.I. and Nilsson, S. (1998) *Journal of Chromatography. A*, **817**, 5–13.

56 Andersson, L.I. (2000) *Journal of Chromatography B*, **739**, 163–173.

57 Wulff, G. (2002) *Chemical Reviews*, **102**, 1–27.

58 Alexander, C. Davidson, L. and Hayes, W. (2003) *Tetrahedron*, **59**, 2025–2057.

59 Piletsky, S.A., Panasyuk, T.L., Piletskaya, E.V. *et al.* (1999) *Journal of Membrane Science*, **157**, 263–278.

60 Jenkins, A.L., Uy, M.O. and Murray, G.M. (1999) *Analytical Chemistry*, **71**, 373–378.

61 Ostroski, A.S. and Stambaugh, R.B. (1950) *Journal of Applied Physics*, **21**, 478–482.

62 Price, G.J. (1996) *Ultrasonics Sonochemistry*, **3**, 229–238.

63 Price, G.J. (2003) *Ultrasonics Sonochemistry*, **10**, 277–283.

64 Biggs, S. and Grieser, F. (1995) *Macromolecules*, **28**, 4877–4882.

65 Svenson, J. (2006) *Analytical Letters*, **39**, 2749–2760.

Index

a
accelerated solvent extraction (ASE) 56, 57
acetonitrile (ACN) 112
acetophenone 134
– aldol reactions of 134
acid extractions 37
acoustic streaming 81, 84, 88
adsorptive stripping voltammetry (AdSV) 89
agitation system 71, 72
alkaline extractions 36
alkylation steps 111, 112, 117
Allium sativum 34
alumina particles 84
– suspension of 84
2-amino-1,4-naphthoquinones 131
– synthesis of 131
amylases hydrolyze starch 41
analytes 72
– solid–liquid extraction of 72
analytical chemistry 6, 129
– common ultrasonic devices used in 6
analytical minimalism 14
analytical techniques 139
anodic stripping voltammetry (ASV) 89
aqueous calibration 30
aqueous media 11
aqueous non-electroactive electrolyte 95
– behavior of 95
arsenic speciation 34
arsenobetaine (AsB) 43
arsenochlorine (AsC) 43
atomic absorption spectroscopy (AAS) 17, 89
atomic emission detector 34
atomic fluorescence spectrometry (AFS) 29

b
barium 24
biphasic analysis 92

biphasic Sonoelectroanalysis 90
blood 92
– detection of copper in 92
boron-doped diamond (BDD) electrode 89
Brassica juncea 34
bubble dynamics 93
bubbled gas 5

c
C_{18} beads 122
carbon-based materials 97
cathodic stripping voltammetry (CSV) 89
cavitation 1, 3, 14, 15, 82, 88
centre–centre distance 100
chalcogenides 138
chelating agent 27
chemical vapor formation 49
chicken tissue 38
– speciation in 38
chlorinated pesticides 72
chromatographic speciation approaches 38
clean-up method 70
closed system 46
C–O bond breaking 145
cold vapor (CV) generation 28
common analytical techniques 17
Coomassie blue method 110, 117
correct ultrasonic probe 12
cross-coupling reaction 131
cup horns 14, 15
current–time responses 93
current ultrasonic devices 14

d
dead cavitation zone 12
dead zone 11, 114
decoupling 13
dedicated vessels 13

Ultrasound in Chemistry: Analytical Applications. Edited by José-Luis Capelo-Martínez
Copyright © 2009 WILEY-VCH Verlag GmbH & Co. KGaA, Weinheim
ISBN: 978-3-527-31934-3

desalting procedure 122
desorption process 73
– facilitating 73
Desulfovibrio desulfuricans G20 112
Desulfuvibrio desulfuricans ATCC 27774 112
Desulfuvibrio gigas NCIB 9332 112
2D-GE separation 110
dichloromethane 64
differential pulse voltammetry (DPV) 90
diffusion layer model 81
2,5-dihydroxybenzoic acid (DHB) 143
dimethyarsinic acid (DMA) 34, 35, 43
direct SPME 71
– improving extraction procedure in 71
dithiothreitol (DTT) 110, 112
dogfish liver 36
Dowex resin 47
Dr Hielscher company 12
dried droplet method 142
drug delivery 113
– ultrasonic applications in 113

e

ECE-type reactions 84
electroactive species 100
electrochemical cell 87
electrochemically reversible ferrocene voltammetry 88
electrode-electrolyte interface 98
electromagnetic stirring 138
electrophoresis applications 59
electrospray ionization (ESI) 121
electrospray sample deposition 142
electrosynthesis 86
electrothermal atomic absorption spectrometry (ET-AAS) 21
element extraction protocol 17
element speciation 19, 30
– extracting reagents for 19
endothermic reactions 8
environmental conditions 19
– changes in 19
environmentally friendly sonochemical method 138
enzymatic digestions 117
enzymatic hydrolysis 39
enzyme ageing 42
enzyme–substrate system 41
equilibrium temperature 9
ET-AAS technique 28
external pressure 5
extracting reagent 17

extraction device 73
extraction techniques 55
extraction temperature 63
extraction time 63

f

F-AAS technique 28
face-on mode 87
Faraday constant 81
fatty acids methyl esters (FAMES) 63
ferricyanide redox 84
– sonovoltammetric response for 84
ferrocene, *see* electroactive species
Fourier-transform infrared spectroscopy 139
fractionation 20, 31
frequency 63

g

gas chromatography (GC) 34, 59
gas–liquid separator (GLS) 29
gel permeation chromatography 139, 140
glassy carbon (GC) 97
gold nanoparticles 138
– preparation of 138
graphite particles 97
– interaction of 97
green energy 129

h

halobenzophenones 84
– dehalogenation of 84
headspace analysis (HS) 56, 58
– dynamic 56, 58
– static 56, 58
heating 117
high performance liquid chromatography (HPLC) 59
high-intensity ultrasound irradiation 134
high-power sonication 72
high-throughput applications 15
high-voltage electrical field 142
homocoupling reaction 131
homogeneous sonochemistry 84
homogenization process 143
honey aroma compounds 75
hot-spot theory 83
HS-SPME 73
– improving extraction procedure in 73
human urine, *see* toxic arsenic
hydride-forming elements 29
hydrodistillation (HD) 75
hydrodynamic electrode 88

hydrodynamic techniques   89
hydroxyl radicals   3, 86

i

ice bath strategy   13
imines   130
– preparation of   130
imprinted polymers   145
– preparation of   145
in situ derivatization   36
indirect ultrasonic applications   47
inductively coupled plasma-mass spectrometry (ICP-MS)   89
in-gel digestion process   108, 110, 111,115
inorganic mercury   29, 37
inorganic nanomaterials   137
– ultrasonic enhanced synthesis of   137
in-solution digestion process   108, 126
in-solution protein   122
– alkylation   122
– reduction   122
in-solution protein denaturation   121
in-solution trapping agent, *see* chelating agent
interfacial cavitation   95
– investigations into   95
International Union for Pure and Applied Chemistry (IUPAC)   30
iodoacetamide (IAA)   110, 112
ionic liquids (IL)   136
– drawbacks of   136
– preparation of   136
isoelectric focusing (IEF) gel strips   110
isotope dilution inductively coupled plasma mass spectrometry (ID-ICP-MS)   34

k

Kimchi   73

l

labor-intense handling   111
α-lactalbumin   124
large-bandgap semiconductor   139
limit of detection (LoD)   89
linear sweep voltammetry (LSV)   89
lipases hydrolyze fats   41
liquid jets   113
liquid samples   43, 64
– speciation in   43
liquid–heterogeneous system   138
liquid–liquid extraction (LLE)   48, 56
– accelerating   48
liquid–liquid interface   84, 86
liquid–liquid partitioning (LLP)   70

liquid–solid element separation   45
– resin for   45
lithium amide   134
– interaction of   134
lobster hepatopancreas   36
low acid concentrations   24
luminescent nanoparticles   138

m

macrocyclic polyamines   131
– chemistry of   131
magnetic stirring   38
Mannich-like reactions   131
mass-based spectrometry techniques   108
– protein identification   108
mass spectrometry (MS)   108
mass spectrometry-based techniques   139, 140
mass spectrometry measurements   117
matrix   20
– mass of   20
– type of   20
matrix-assisted laser desorption/ionization (MALDI)   141
matrix–polymer couple   141
matrix solid phase dispersion (MSPD)   75
mechanical stirring   131
meso-cyanophenylporphyrins   134
– interaction of   134
methyl salicylate   131
– hydrazinolysis of   131
methylarsonic acid (MMA)   34
methyl-mercury   36
Michaelis constant (K_m)   41
microelectrodes arrays   93, 94
microjetting   84
microplate horn   15
microreactor   3
micro-sized irregular polyanilines   131
– synthesis of   131
microsyringes   113
microwave assisted extraction (MAE)   56, 58
microwave energy   35, 38, 117
– use of   38
modern ultrasonic probes   6
molecular imprinting method   145
molecular weight (MW)   110
molecularly imprinted polymer (MIP)   145
monoatomic gases   6
monomers   139
monomethylarsonic acid (MMA)   34
multiple probes   11
mussel tissue   20, 41

n

nanoparticles   137
nanoplates   137
nanotubes   137
N-benzoyl-N-phenylhydroxylamine   92
Nernst diffusion layer   81
Nernst diffusion model   94
– application of   94
Nernst model   82
nitrobenzenes   84
– dehalogenation of   84
non clean   122
non-accelerated protocol   122
non-buffered media   41
non-electroactive solutions   100
non-enzymatic approaches   36
non-particle metal intrusion   11
nonvolatile organic compounds   58
nuclear magnetic resonance   139
number-average molecular weight (M_n)   140

o

on-line applications   45
on-line ultrasonication   48
– coupling of   48
open system   46
optical storage media   138
optoelectronic devices   137
organic compounds   56, 70
– extraction methods used for   56
– modern extraction techniques   56
– optimum extraction efficiencies for   70
organic mercury species   44
organic solvents   19
– use of   19
organic synthesis   129
– sonochemistry for   129
organochlorine pesticides (OCPs)   57
organo-phosphorus pesticides (OPPs)   58
Oryza sativa   35
oxidizing agents   18

p

PAH extraction   12
Parmesan cheese   73
– volatile compounds of   73
particle impact experiments   96
particle–solution contact   14
peak tailing   147
peptide mass fingerprint (PMF) approach   108
– disadvantage of   108
– protein identification through   108
– sample treatment for   108
– in-solution protein sample treatment for   118
petrol   90
– determination of lead in   90
photochemical reactions   136
– effect of ultrasound on   136
photorefractive devices   138
photovoltaic solar cells   138
plants   24, 34
– speciation from   34
– US-SLE extraction from   24
plume   141
PMFmethodology   124
– major drawback   124
poly(ethylene glycol) (PEG)   143
– analysis of   143
– ultrasonic scission of   145
polychlorinated biphenyls (PCBs)   57
polycyclic aromatic hydrocarbons (PAHs)   57
polydimethylsiloxane (PDMS) liquid phase   59
poly-dispersity-based techniques   139
polydispersity index (PDI)   140
polyether ether ketone (PEEK)   131
polymer characterization   140, 141, 145
– ultrasonic-induced polymer degradation for   145
polymer degradation mechanisms   139
polymerization   131
polymers   139, 141
– characterization of   139
– MALDI analysis of   141
– properties of   139
– structural analysis of   141
polystyrene (PS)   143
polytetrafluoroethylene (PTFE)   59, 131
potentials of zero charge (PZC)   95, 96, 97
– measurement of   95
power magnification   10
power ultrasound   81
– electrochemical applications of   81
pressurized liquid extraction (PLE)   56
protein–acetonitrile interactions   125
protein biomarkers   118
protein depletion   118
protein digestion   108, 111, 121
– strategies of   108
protein identification   116, 121
– using ultrasonic protocol   116
protein staining   117
– influence of   117
protein visualization   110

– modern stains for 110
proteins 5
– enzymatic digestion of 5
proteome, *see* proteomics
proteomics 107
– challenge of 107
purge-and-trap (PT) 56

q
quart cell (Q) 29

r
Raman spectroscopy 139
reaction container 9, 14
– material of 9, 14
– shape of 9, 14
reagent concentrations 18
refrigeration 15
reverse phase ion pair (RP-IP) 35
reversed-phase HPLC 141

s
sample handling 113
Sample Preparation Techniques in Analytical Chemistry, the book 56
Se-amino acids 35
Se speciation 39
search engines 108
sedimentation velocity 140
selenium-enriched lentil plants 35
sequential enzymatic extraction 42
sequential extraction schemes (SES) 19, 20
– extracting reagents for 19
sequential injection/flow injection analysis (SIA/FIA) system 48
Se-species 35
sewage sludge 24
– US-SLE from 24
shortening sequential fractionation schemes 31
shotgun proteomics 118
shot-to-shot analysis 142
Shoup–Szabo approximation 100
signal-to-noise ratio 116, 117
silica ultrasonic probe 11
silver nitrate methods 110
single-crystalline Cd(OH)$_2$ nanorings 138
– sonochemical fabrication of 138
small-angle neutron scattering (SANS) 139, 140
small-scale ultrasound-assisted extraction procedure 20
sodium dodecyl sulfate (SDS) 110
soft tissues 27, 36

– extraction from 27
– speciation from 36
soils and sediments 24
– US-SLE from 24
sol–gel reactions 137
solid phase extraction (SPE) 56
solid phase microextraction (SPME) 56, 59, 71
– coupling 71
solid powders 28
solid samples 64
solid–liquid elemental extractability 12
solid–liquid extraction process 6, 12
– of metals 4
  – efficiency of 4
solid–liquid interface 83, 85
solid–liquid partitioning (SLP) 56
solute–matrix interactions 5
solvent extraction 92
solvent-free extraction technique 59
solvent-free sample preparation 142
solvent temperature 5
solvent volume 62
sonic power 62
sonication 4, 23, 36, 72, 73
– intensity of 4
– temperature of 23
sonication amplitude 115
sonication conditions 32
sonication frequency 23
– importance of 23
sonication time 115
sonication volume 115
sono-anodic stripping voltammetry (Sono-ASV) method 90
– applications of 90
– validity of 90
sonochemical degradation 61
sonoreactor (SR) 6, 14, 22, 114, 142
sonotrode 87
sonovoltammetry 84, 87
sonovoltammogram 88
– appearance of 88
Soxhlet extraction 56
spiral probes 11
spot-to-spot analysis 142
stable cavitation 83
steady-state voltammetric response 93
stir bar sorptive extraction (SBSE) 56, 59, 74
– coupling 74
– ultrasonic treatment in 74
stripping voltammetry 88
– trace detection by 88

strong anion-exchange solid-phase extraction (SAE-SPE) 44
sulfate-reducing bacteria 112
supercritical fluid extraction (SFE) 56, 57
surface-active impurities 89
– real-world samples of 89
Suzuki reaction 131
synthetic polymers 139
synthetic procedures 129
– inorganic 129
– organic 129
synthetic solids 20
sypro orange fluorescent dyes 12
sypro red dyes 12

t

temperature 116
temperature control 13
Tessier protocol 32
tetraazacycloalkanes 131
– synthesis of 131
tetrasubstituted porphyrin 134
– schematic reaction for 134
three-electrode electrochemical cell 87
time-resolved chronoamperometric signals 96
– study of 96
titanium alloy 10, 11
titanium dioxide 139
titanium probes 12
total element extraction 18
– extracting reagents for 18
toxic arsenic 19, 43
traditional probes 11
transient bubble 3
transient cavitation 83–84
transport-limited electrolysis 94
tributyltin (TBT) 36
tris(hydroxymethyl)aminomethane (THAM) 37
trypsin ratio 115
turbulent convective flow 88
two-dimensional geometric model 100
two-electrode array 100

u

ultrasonic amplitude 23
ultrasonic application 6, 13
– direct 6
– indirect 6
– pulse mode of 13
ultrasonic assisted enzymatic digestion (USAED) 39
ultrasonic assisted extraction (UAE) 55
– basic principles 60
– parameters influencing 61
ultrasonic bath (UB) 6, 7, 9, 18, 45, 72, 142
– types of 7
– ultrasonic intensity distribution 8
ultrasonic cavitation 3
– parameters affecting 3
ultrasonic devices 1, 22, 62
– type of 22
ultrasonic energy 1, 3, 17, 36, 39, 43, 117, 130, 142
– applications of 17
– chemical effects of 3
– effect of 117
– enzymatic kinetics by 39
– mechanical effects of 3
ultrasonic extraction 20, 63
ultrasonic frequency 3, 23
ultrasonic intensities 130
ultrasonic probe (UP) 9, 18, 142
– parts of 10
– types of 10
ultrasonic probe devices 124
ultrasonic radiation 60
ultrasonic slurry sampling (US-SS) 28
ultrasonic solvent extraction (USE) 55
ultrasonic technology 1
– state-of-the-art of 1
ultrasonic wave 9
ultrasonically-enhanced mass transport 84
ultrasonication 20, 23, 43, 139
– advantages of 20
– time of 23
ultrasonication aids 3
ultrasound (US) 1, 70, 71, 84, 86, 90, 93, 129
– applications of 129
– depassivating effect of 85
– electroanalysis facilitated by 90
– electrochemical processes 84
– power of 1
ultrasound assisted extraction (UAE) 76
ultrasound operational parameters 61
– optimization of 61
unambiguous protein identification 124
uniformly-accessible electrode 81
United States Environmental Protection Agency (USEPA) 11, 55
UP 47
US-induced agglomeration 139

v

vanillin 92
– determination of 92
– extraction of 92

vapor pressure osmometry   140
volatile organic compounds (VOCs)   56
voltammetry   87
– under insonation   87
vortex mixing   142

w

wall-tube electrode   84
washing procedure   112
weight-average molecular weight (M_w)   140

wet grinding methods, *see* solvent-free sample preparation
Williamson ether synthesis   131

x

X-ray scattering   139
– small-angle   139
– wide-angle   139

z

ZipTips   125
Znq_2 nuclei   138
– nucleation of   138